Building the 21st Century City through Public–Private Partnerships

Building the 21st Century City through Public–Private Partnerships introduces students and early-career professionals to the fundamentals of this unique form of cross-sector collaboration. From understanding the responsibilities of government and industry partners to stewardship of taxpayer dollars, this introductory guide empowers developers and local officials to deliver successful commercial, leisure, and industrial projects neither could undertake on their own. Chapters on securing financing and navigating permitting processes demystify the steps to creating profitable developments, while case studies from around the United States provide invaluable local context. A glossary of public–private partnership terminology offers the reader an insider's grasp of the language of government and industry partnerships.

- Equips developers and local officials with the foundations for successful collaboration
- Provides a template for building effective public–private partnerships in every area of real estate development
- Includes field-tested insights from case studies of diverse public–private partnership examples
- Ideal reading for courses in public administration, city planning, real estate, not-for-profit studies, public service, and more

Helmed by a practitioner turned academic, *Building the 21st Century City through Public–Private Partnerships* serves as a masterclass with veteran developers, planners, municipal officials, and scholars.

Stephen T. Buckman is an Associate Professor of Real Estate Development at Clemson University. He is 2023/2024 Fulbright Scholar award recipient to Brazil where he will study land value capture and its impacts on real estate in Rio de Janeiro. His research is primarily centered on resiliency, waterfront development, community real estate development, and how public–private partnerships bring these all together. Before entering the academy, Dr. Buckman worked in real estate development for the Trammell Crow Investment Division and had his own small firm as well as working in investment commercial brokerage. At present Dr. Buckman consults for developers and communities and does his own multi-family developments. Dr. Buckman holds a PhD from Arizona State University.

Building the 21st Century City through Public–Private Partnerships

A Tool for Real Estate Development and Urban Growth

Edited by Stephen T. Buckman

NEW YORK AND LONDON

Designed cover image: Stephen T. Buckman

First published 2024
by Routledge
605 Third Avenue, New York, NY 10158

and by Routledge
4 Park Square, Milton Park, Abingdon, Oxon, OX14 4RN

Routledge is an imprint of the Taylor & Francis Group, an informa business

Library of Congress Cataloging-in-Publication Data
Names: Buckman, Stephen, editor.
Title: Building the 21st century city through public-private partnerships / edited by Stephen Buckman.
Description: First Edition. | New York, NY : Routledge, 2024. | Includes bibliographical references and index.
Identifiers: LCCN 2023011580 (print) | LCCN 2023011581 (ebook) | ISBN 9781032120829 (hardback) | ISBN 9781032120690 (paperback) | ISBN 9781003222934 (ebook)
Subjects: LCSH: Cities and towns—Growth. | Public-private sector cooperation. | Public-private sector cooperation—Finance. | Real estate development.
Classification: LCC HT371 .B775 2024 (print) | LCC HT371 (ebook) | DDC 307.76—dc23/eng/20230414
LC record available at https://lccn.loc.gov/2023011580
LC ebook record available at https://lccn.loc.gov/2023011581

ISBN: 978-1-032-12082-9 (hbk)
ISBN: 978-1-032-12069-0 (pbk)
ISBN: 978-1-003-22293-4 (ebk)

DOI: 10.1201/9781003222934

Typeset in Times New Roman
by Apex CoVantage, LLC

Contents

Contributor Biographies *vii*
Foreword *x*
JOHN TALMAGE

SECTION 1
An Overview of Public–Private Partnership and Its Tools 1

1 **An Introduction to Public–Private Partnerships as a Development Paradigm** 3
STEPHEN T. BUCKMAN

2 **P3 Tools: TIF, BIDs, Brownfields, and Eminent Domain** 14
JEFF BURTON

3 **Contracts: The Heart of Public–Private Partnerships** 28
REED L. BENNETT

4 **How Cities Use PPPs to Spur Real Estate Development: A Look at Greenville, South Carolina** 39
NANCY P. WHITWORTH

5 **Measuring the Costs and Benefits of a TIF-based Public–Private Partnership** 61
JEFF BURTON

SECTION 2
Public–Private Partnerships in Action 81

6 The Use of Public–Private Partnerships to Redevelop Greenville's West End through a Minor League Baseball Stadium Development 83
STEPHEN T. BUCKMAN AND JAMES FRAZIER

7 Innovation Districts and Misplaced Economic Development Incentives 95
CARLA MARIA KAYANAN AND PATRICK COOPER-MCCANN

8 Industrial Development and Public–Private Partnership: The Enigma Case Study 112
BROCKTON HALL

9 Partnering with Public Agencies to Revitalize Blighted Areas 127
JOSEPH BONORA

10 Leveraging P3 to Increase the Financial Upside of the Navy's Underutilized Real Estate: A Case Study at Naval Air Station Oceana, VA 136
MICHAEL YEARY AND STEPHEN T. BUCKMAN

Appendix *151*
Glossary of Key P3 Terms *175*
References *184*
Index *193*

Contributor Biographies

Reed L. Bennett recently completed a Master of Real Estate Development degree at Clemson University and is a licensed transactional attorney in Georgia. He has served as counsel on municipal bond issuances of more than $3 billion dollars over a 3-year period while practicing with the law firm Kutak Rock LLP, in Atlanta, GA. These bond issuances financed a range of educational, mass transit, multi-family, and other governmental projects. Responsibilities while at Kutak Rock LLP consisted of drafting transactional documents; preparing offering documents and other disclosure materials related to the sale and issuance of municipal securities; reviewing and analyzing various due diligence materials, contracts, and third-party reports to ensure accurate disclosure to potential investors; attending public meetings and hearings; mentoring less experienced lawyers; and serving as the only associate attorney on the Advisory Committee for the firm's Atlanta Office.

Joseph Bonora is the founder and President of Catalyst Asset Management, Inc. and Catalyst Community Capital, Inc. Joe has over 20 years of experience in real estate finance and development, and is responsible for managing the strategic direction of the companies and identifying new real estate development and investment opportunities. Joe is particularly well versed in government funding and tax credit programs, and is well known for his ability to engineer complex capital structures and create innovative new ways to finance projects. Prior to forming Catalyst, Joe was the co-founder and Managing Director of Aileron Investment Management, a real estate development and specialty finance firm focused on the development of multi-family, senior housing, and mixed-use properties, and the origination of government guaranteed loans. From the time he formed the company in 2010 until his departure in 2018 to start Catalyst, Joe led Aileron in providing over $660 million in small business and commercial real estate loans nationwide, and developing over $200 million in multi-family and commercial real estate projects in Florida.

Stephen T. Buckman (editor) is an Associate Professor of Real Estate Development at Clemson University. He is 2023/2024 Fulbright Scholar award recipient to Brazil where he will study land value capture and its impacts on real estate in Rio de Janeiro. His research is primarily centered on resiliency, waterfront development, community real estate development, and how public–private partnerships bring these all together. Before entering the academy, Dr. Buckman worked in real estate development for the Trammell Crow Investment Division and had his own small firm as well as working in investment commercial brokerage. At present Dr. Buckman consults for developers and communities and does his own multi-family developments. Dr. Buckman holds a PhD from Arizona State University.

Jeff Burton is the City of Tampa, Florida, Downtown Community Redevelopment Agency Director. Dr. Burton is an experienced State of Florida licensed residential building contractor and has performed over 5,000 building safety inspections as a licensed building code administrator. While at the Insurance Institute for Building and Home Safety (IBHS) he coauthored the Louisiana State building code law in the aftermath of Hurricane Katrina and was a founding research member of the State of Florida post-Hurricane Charley building code analysis team. Dr. Burton is also the incoming President of the Florida Redevelopment Association (FRA). Dr. Burton holds a PhD from the University of South Florida.

Patrick Cooper-McCann is an Assistant Professor of Urban Studies and Planning at Wayne State University. His work is centered on the redevelopment of the Post-Industrial City through private-public partnerships with a particular interest in the Detroit region.

James Frazier currently works in acquisitions for RealOp Investments, a commercial real estate investment and private equity firm located in Greenville, SC. He has over 10 years of experience in commercial real estate research, reporting, and transactions, and holds a Master's of Real Estate Development from Clemson University and a BA in Economics from the University of North Carolina at Chapel Hill.

Brockton Hall is a Vice President with Colliers International, specializing in industrial brokerage. Mr. Hall concentrates most of his work in the Carolinas but also consults to clients in helping with location choices throughout the United States. Mr. Hall holds an MRED in real estate development and a BS in Construction Management from Clemson University.

Carla Maria Kayanan is a political-economic geographer with strong interests in the spatial organization of work in the tech-economy and the resultant landscapes of urban inequality. She is an Assistant Lecturer at the Maynooth University Social Sciences Institute in Ireland. Dr. Kayanan holds a PhD from the University of Michigan, an MA from the University of Chicago, and a BA from the University of Maryland College Park.

John Talmage is the Director of Lee County Office of Economic Development and real estate developer. He is responsible for attracting $5 billion of new development to Lee County, ranging from affordable housing to corporation headquarter attraction to new community revitalization initiatives.

Nancy P. Whitworth is a private economic development consultant. She was previously the Director of Economic Development and Deputy City Manager for the City of Greenville and was responsible for: commercial and neighborhood revitalization; downtown development; business recruitment and retention; planning and zoning; and building codes. Whitworth has served as an expert panelist and lecturer throughout the country, advising communities on downtown revitalization, strategic planning, and public–private partnerships.

Mike Yeary is a Lieutenant Commander in the United States Navy's Civil Engineer Corps and a registered Architect in the state of South Carolina. He currently serves as the Officer in Charge of Construction at Naval Air Station Pensacola, the Navy's premiere aviation training military installation worldwide. He has worked with Naval Facilities for 10 years at large installations in Europe and the United States, delivering over $240 million of work in place for Navy Facilities. He holds a Master's degree in Architecture from the Georgia Institute of Technology, a Master's in City Planning from Clemson University, and is a certified Contracting Officer from the DoD's Acquisition University. He is a member of the Society of American Engineers (SAME), the Institute of Classical Architecture, and the Congress for New Urbanism.

Foreword

John Talmage

How often have Americans woken up to news that a bridge has collapsed, such as the Silver Bridge connecting West Virginia and Ohio killing 46, or a municipal water system has failed, and thousands of residents are put on a boil notice for months, such as in Flint Michigan and Jackson Mississippi? It is estimated by the American Society of Civil Engineers that it would cost nearly $2.6 trillion over ten years to bring our current infrastructure into good repair. In another report, the American Road and Transportation Builders Association highlight that of the United States' 620,000 bridges, one third need major work or replacement. How this infrastructure crisis gets financed and who pays for it raises complex issues for municipal finance experts, municipal and state governments who may or may not have the expertise to address these issues, and citizens who may question who owns their highway or courthouse. Just this month, the Florida Tax Watch has proposed new public–private partnership regulations and warns that if the Florida Legislature does nothing then by year 2039, a continued underinvestment in Florida's infrastructure at current rates will have serious economic consequences—$10 trillion in lost Gross Domestic Product (GDP), more than 3 million lost jobs, and $2.4 trillion in lost exports. This book seeks to raise many of these questions and uses multiple examples of how a public–private partnership can work and serve as an effective tool for local governments specifically to begin much-needed infrastructure involvement.

A public–private partnership (PPP) is a long-term arrangement between a government and the private sector. Typically, it involves private capital financing government projects to design, build, and often to operate the project and then taking revenue from either the project or other government sources to cover the cost of the project with an adequate return over the course of the PPP contract. These projects often work best when supported by an independent revenue source such as a water rate, a toll, or a user's fee. Revenue tied to financing is the most basic form of PPP, but certainly not the only model.

There are various PPP contract models based on funding, the partners of which are responsible for owning and maintaining assets at different stages of the project. This arrangement is often thought of as a continuum depending on how much involvement the public sector has versus how much involvement the private sector has. Whatever model is chosen, the contract must capture how performance will be measured by each participant so that accountability is insured.

Examples of PPP models pulled from the literature include:

- Design-build (DB). The private-sector partner designs and builds the infrastructure to meet the public-sector stakeholder's specifications, often for a fixed price.
- Operation and maintenance contract (O&M). The private firm, under contract, operates a publicly owned asset for a specific period. The public partner then retains ownership of the assets.
- Design-build-finance-operate (DBFO). The private-sector company designs, finances, and constructs a new infrastructure component and owns the operation and maintenance under a long-term lease. When the lease is up, the private-sector partner transfers the infrastructure component to the public-sector partner.
- Build-own-operate (BOO). The private party finances, builds, owns, and operates the infrastructure component perpetually. The public-sector partner's constraints are stated in the original agreement and through ongoing regulatory authority.
- Build-own-operate-transfer (BOOT). Privatization is granted for financing, design, building, and operation of an infrastructure component (and to charge user fees) for a specific time, after which ownership is transferred back to the public-sector partner.
- Buy-build-operate (BBO). This publicly owned asset is legally transferred to a private-sector partner for a designated period.
- Build-lease-operate-transfer (BLOT). The private-sector partner designs, finances, and builds a facility on leased public land. The private-sector partner operates the facility for the duration of the land lease. When the lease expires, assets are transferred to the public-sector partner.
- Energy Service Companies (ESCO). The company provides energy efficient technology to reduce energy costs whose savings are used to support the financing.

There are benefits and drawbacks to the PPP model. The private sector is often in a better position to design and build a project faster and more efficiently. It can include its understanding of best practices and innovation as a subject matter expert in the sector. Risk is transferred from the public sector and public funds can be freed up. While these are all considered valuable, there are worries that the public sector may not understand the complexity of the deal. Projects that are scheduled to last a long time may face cost overruns and scheduling delays, and the public sector may take on more debt than it has the long-term ability to repay.

Whatever model is used, they have emerged over an extensive period in the United States and around the world. After all, all levels of government use public–private partnership to build public infrastructure. Today, these projects range from privatized prisons to utilities to roadways and bridges. While some of the asset classes may be new, such as broadband, the use of the private sector to finance, construct, and manage public services goes back for centuries, and in the United States to the earliest days of the country's founding. Private parties' involvement in highway projects goes back to the late 1700s, when the first turnpike was built—The Philadelphia & Lancaster, the first hard-surfaced road. The Commonwealth of Massachusetts could not afford to pay for its construction, so it was privately built. In 1785, the Charles River Bridge Company requested the state legislature to grant it the right to build a bridge across the Charles River to connect Cambridge and Boston. The bridge's proponents argued that the two towns had grown, and that the existing ferry was neither efficient nor effective moving traffic across the river. The Charles River Bridge Company was given the right to build a bridge and collect tolls for 40 years. In the 1800s, mail was carried between physical post office locations by private mail contractors. This eventually became the Pony Express. The Pacific Railroad Act of 1862 gave railroad companies five square miles of land on each side of the track for every mile of track laid, which they sold subsequently as the land became valuable.

Over the past 20 years, public–private partnerships in the United States have increased substantially. This has prompted states to begin adopting regulatory frameworks to address these projects so that there is some transparency, and municipalities understand what their responsibilities are. The United States faces a growing need to build and maintain critical infrastructure—everything from airports to wastewater treatment plants—but with limited government funding to do the job. Encouraging private sector investment in infrastructure is an essential part of the solution. Whether it is infrastructure banks, federal guarantees, or municipal bonds, there is no one single strategy that can adequately fund this gargantuan bill coming soon. Every cracked bridge, failed highway, or crumbling water system is a constant reminder that we have not adequately maintained what we have even though we continue to build additional capacity. One area that could leverage substantial private investment—public–private partnerships—is currently limited or unavailable in most states due to lack of enabling legislation. Thirty-three states (along with the District of Columbia and Puerto Rico) have enacted by statute some sort of PPP-enabling legislation. These legislative initiatives and each state's approach to PPP have created a complex investment environment that is met with varying degrees of success and public support.

Advances in technology and digital ecosystems, vendor lock-ins, and other impediments that may make it too costly to change a technological direction after the investment has been made also make some PPP projects difficult. Although we may need to consider the constraints that technological advances may place on a project, we may very well see great strides in addressing climate change and other unknown challenges to a PPP project. These are all factors that must be worked into a contract in an appropriate way.

Now that we are seeing firsthand the impact of severe weather and other manifestations of climate change, adequate and properly functioning infrastructure will be more critical than ever to insure our wellbeing and future quality of life. Replacing what is at risk or has exhausted its shelf life while designing new infrastructure to meet future challenges is a very tall order. To do this with adequate transparency and accountability makes it even more difficult, but this is the time we need to find the right models to build partnerships that can meet this difficult task.

Section 1

An Overview of Public–Private Partnership and Its Tools

1 An Introduction to Public–Private Partnerships as a Development Paradigm

Stephen T. Buckman

Introduction

This book is a product of necessity. As an academic and real estate development practitioner who teaches a public–private partnership (P3) course at the graduate level, I found it difficult to find adequate reading material for my students when it came to P3 and the real estate sector. To be sure, there is a vast body of literature, both in article form and as book manuscripts, that discusses the relationship between P3 and public infrastructures, ranging from the building of highways to the building of prisons. However, with the exception of John Stainback's now dated book, *Public/Private Finance and Development* and Chris Nelson's book, *Foundations of Real Estate Development Financing: A Guide to Private-Public Partnerships*, which deals mostly with the numbers behind P3, there is surprisingly little out there that deals specifically with the real estate development process beyond financing, hence the need for this edited volume on this missing subject.

P3 comes in many forms, including the development of public infrastructure with which it is traditionally associated and the community real estate development where it is mostly associated with efforts to reconfigure urban landscapes. The heart of P3, however, is embedded in its name, "partnership"; specifically, a partnership between "private" and "public" sectors, with each relying on and complementing the other's strengths to deliver successful projects. Typically, the private sector contributes a long history of development and innovation from numerous contractors to financing to the latest trends in design and development. At the same time, while the private sector has these and many other arrows in its quiver, it needs help streamlining development projects and managing the many complexities that are involved in the entitlement and predevelopment processes, which is expertise that is far more prevalent in the public sector. However, while the public sector is adept at streamlining and navigating the bureaucratic red tape that is associated with cities' entitlement processes, it often lacks the practical expertise of organizing and managing a development all the way from financing to bricks and sticks to the completed development. Thus, each entity in the development process is relying on its partnerships to coordinate their respective skillsets in order to increase the timely completion and success of real estate development projects.

DOI: 10.1201/9781003222934-2

Understanding the above nuances and how actual partnerships work has become a necessity for those on both sides of the real estate world. Traditionally, there has been an artificial dichotomy between the two sectors within the development process. The private sector has often blamed municipalities for hampering their ability to develop by enforcing outdated zoning, permits, and overall red tape. On the flip side, municipal building, planning, and zoning departments have often treated developers as pillagers of the land who, without some regulatory speed bumps such as zoning and permitting, would run roughshod over land markets. The end result, according to this view, is the production of urban landscapes that serve overwhelmingly developers' return on investment (ROI) at the expense of overall community development.

While this mutually antagonistic view dominated real estate development thought, and continues to hold sway in far too many circles, the pendulum appears to be swinging to a point where it is easier and often more profitable for public and private interests to be partners rather than enemies. Partnerships are based on mutual needs and desires. For the public sector, there has been a shift in what they perceive their role to be. Traditionally, when it came to real estate development, the role of the public sector was to ensure a balance between growth that was good for the economy and society, but at the same time minimize harm to the community. As mentioned earlier, this task was achieved through regulatory speed bumps on rampant development. But a sea change has occurred in many urban governments over the last few decades. Rather than hindering development, the public sector has itself become a developer. Nowhere is this sea change more apparent than in urban housing authorities, which almost universally have shifted from their original social mission of housing poor and working-class people to becoming developers in their own right. As such, these housing authorities are not only concerned with providing affordable housing but also with achieving solid ROIs. With the rebirth of urban business districts, a growing number of urban governments over the last few decades have been pressured to become part of the urban renewal process if they want to reinvigorate their increasingly derelict built environments. In this context, the ideologically dominant way to reinvest in their futures was to fully embrace the development process.

Thus, the public sector turned to partnerships with the private sector to recreate existing built environments. Conversely, the private sector embraced the public sector as a way to increase its lagging ROI and at the same time direct much needed tax revenues to urban governments as their equity partners. This was especially true during the early twenty-first century (2008–2019), when capital markets squeezed developers for increasingly bigger ROIs, and continues to be an issue with the onset of recession and interest rate manipulation by the Federal Reserve Bank in 2022 and 2023 to ostensibly curb inflation. In both cases, the financial markets and the Fed squeezed cap rates, making it more difficult for developers to adequately fund urban development projects. Other pressures on developers include holding or carrying costs (typically in the form of loans) that can kill development projects' profit margins, especially when lending requirements are becoming thinner than ever. Moreover, beyond securing financing, the predevelopment/entitlement permitting process also contributes to greater holding costs. For instance, in multi-family development projects, 40 percent of the cost of development derives from public regulations, which include items such as rezoning, site fees, permitting, Occupational Safety and Health Administration (OSHA) regulations, and so on (Emrath & Sugrue Walter, 2022; Shaver, 2022). Thus, partnerships with urban entities have become more appealing as they smooth out the entitlements process and provide incentives, often in the form of tax rebates and development financing.

So, in the present state of real estate development, both the public and private sectors are increasingly discarding the traditional perception that they are enemies and are instead becoming partners seeking mutual gain. This book explores this idea of mutual gain by explaining how P3s are structured and providing examples of this development paradigm in action. The goal of this book is for readers to get a better theoretical understanding of the putative benefits of P3 for both the public and private sectors and how they might apply some of the examples from the book to their own developments or partnerships. The next section of this introduction briefly outlines the foundation of P3 so that the narrative structures and case examples that follow will be more easily comprehended. To that end, this section will first discuss the background of P3, define the terminology, and explain the structures of the rest of the book. Following this overview, I will outline each chapter and its importance to the P3 debate.

Public Partnerships: An Overview

We must remember that P3 by its very nature is a structured and defined partnership. According to Nelson (2014: 4–5): “P3s are contractual relationships between public and private entities to facilitate real estate development . . . that would not occur without one partner or the other.” In this simple definition, there is a clear implication that each partner is bringing something to the table to benefit the other. Expanding on this definition, the Urban Land Institute (ULI), a leading U.S.-based real estate organization, states that P3s are “creative alliances” between a government entity and private developers to achieve a common purpose (Friedman, 2016). So, they are partnerships and alliances, with the key takeaway being that, without either party, any development in question would not get built.

In this light, a strong working definition of P3 in relation to real estate development would be:

> A partnership between public and private entities to achieve a solution, to deliver an infrastructure and/or real estate development solutions over the long term. It combines the strength of the public sector’s mandate to deliver services and its role as a regulator and coordinator of public functions with the private sector’s focus on profitability, ingenuity, malleability, and commercial efficiency.
>
> (Adapted from Stainback, 2002)

Based on this working definition, it becomes important to recognize some of the fundamental characteristics of a P3. First and foremost, a P3 represents an arrangement whereby the private sector is contracted to deliver infrastructure/real estate services to the public or to assist the public in that task. Therefore, for a partnership to exist, it must include these key fundamentals: (i) be contractual, meaning it cannot just be back of napkin; (ii) be focused on delivery of services or performance such as a building or some type of infrastructure; (iii) be long term; (iv) be focused on risk sharing; (v) involve at least one public and one private agency; and (vi) embrace the principle of one size does fit all and simpler is often better.

Understanding these fundamentals, it is important to lay out (i) the motivations of each party; (ii) sources of financing; (iii) the key actors involved; (iv) the policy framework guiding the partnership; and (v) the organization of the agency that will administer and control the P3 process. It is vital to clearly understand each of these key variables before initiating a partnership. The sooner one grasps the contractual obligations, the better, thus allowing all parties to understand what is expected and what the deliverables are.

Motivations of each party. As briefly mentioned earlier, public sector motivations are efficiency throughout the entire asset life solution (What is this? "Asset life solution"); transparency; innovation; know-how; and new sources of financing. For the private sector, motivations include community interaction; zoning; permitting; and financial incentives. Understanding what the exact motivation of each party in the first stage is key. This also fits with the idea of *establishing a policy framework* which would clearly lay out and define the purpose of the P3 and identify the responsibilities of all parties and the conditions and liabilities of each party's participation.

Sources of financing. In essence, each party has access to sources of financing that the other party either cannot get or would have problems obtaining or fully understanding. The public sector can bring sources of funding that are incentive driven through programs like Low Income Housing Tax Credits (LIHTC); Tax Increment Financing (TIF); Historical tax credits, etc. The private sector has access to forms of capital that the public sector cannot access, including both debt and equity. There are essentially four types of financing that could be considered in each P3 deal and often these will be intermingled: (i) Government financing where the government borrows money and provides it to the project through lending, grants, or subsidies; (ii) Corporate financing where a private entity or company borrows money against its proven credit position(s) and ongoing business and invests in the project; (iii) Project financing where non- or limited recourse loans are made directly to a special-purpose vehicle i.e., real estate; and (iv) Private equity where a private individual(s) via their own finances or through investment houses put up money for a project with the guarantee of a specific return.

Key actors in the P3 process. Who then will be the key actors in the P3 partnership? This will be largely dependent on the project and what the goals of the partnership are. But it should be expected that there will be some combination of players. These players could include, but are not limited to: the contracting authority; project company; lenders; offtake purchaser(s) (if the project is to be sold to a third party); input supplier; construction contractor; and, lastly, the operator.

The P3 agency. While it may be more centered on the public side of the equation, this entity is just as important to the private side as it helps to determine the structure of the actors, the financing, and the goals. Thus, it is important to have equal say regarding how the agency is run and who runs it. Remember this: at the end of the day, it is a partnership. In this regard, a solid P3 agency should: improve the policy/regulatory context of the partnership; ensure that the P3 program is integrated within the overall planning, fiscal risk management, and regulatory systems; ensure that the project protects the government, environment, and social interests of the community; and, lastly, promotes P3 opportunities at multiple levels.

Knowing the structure and the underlying needs of both parties for a partnership of this nature, we can see some driving factors or principles that lead to a successful P3. In their 2005 report, ULI (Corrigan et al., 2005) laid out ten principles (Table 1.1) for success, each of which I will explore in more depth.

Table 1.1 ULI's principles for successful public–private partnerships

1. Prepare properly for public–private partnerships
2. Create a shared vision
3. Understand your partners and key players
4. Be clear on the risks and rewards for all parties
5. Establish a clear and rational decision-making process
6. Make sure all parties do their homework
7. Secure consistent and coordinated leadership
8. Communicate early and often
9. Negotiate a fair deal structure
10. Build trust as a core value

1 *Prepare properly for public–private partnerships*: Arguably, this could be the most important part of establishing a partnership and the place where failure occurs. What takes place in this stage lays the foundation for everything that follows. Both parties must prepare for and establish what will result from their partnership. Preparing includes all the factors that will follow below. Preparation also allows both parties to properly weigh the risks and rewards of the partnership. This is also the phase when both parties should flesh out what will be needed to be successful and thus discuss their mutual plans and ideas.

2 *Create a shared vision*: The parties involved must come up with a vision that is shared and communal. It cannot be one party establishing the vision and expecting the other to simply follow. Such a move will result in failure. Rather, there must a be shared vision of not only the results of the project, but also the process to get to the desired results. This vision must include how both parties will succeed in the partnership.

3 *Understand your partners and the key players*: Understanding who your partners and key players are, is key. This goes beyond simply knowing their positions in particular private companies or their positions in government. Understanding involves grasping what drives the respective partners both personally and professionally. For instance, if a mayor or an elected official is leading

a P3 from the public side, their motivations may be very short term, i.e., the election cycle, whereas a Planning Director may be thinking very long term. Each vision, either short or long term, will have an impact on how a development project will get done and what the actual components of that development are. A short-term vision may favor immediate media coverage and political flash, while the long-term vision of, say, a Planning Director, may have a bigger and more comprehensive view of a particular development project.

4 *Be clear on the risks and rewards to all parties*: This point follows the previous point. Depending on who the key players are, what their needs are, and the result they want will determine risks. Taking the example from above, the risks to elected officials may be being voted out of office. It is, therefore, important to know and clearly identify the risks for the private sector versus the public sector. For the private sector, if a deal goes south, while it will hurt financially and there will be some very unhappy equity partners and probably even more unhappy spouses, the pain will be felt by a small group of people. On the other hand, the risk for the public sector has a much greater impact as public development officials may have to answer to the greater community. Thus, if a deal does fail, the public sector will have to potentially cut essential public services such education, streets maintenance, etc. to make up for the financial loss. So, it is necessary that each party identifies its respective risks upfront, which will help to determine the most optimal approach to an actual development project.

5 *Establish a clear and rational decision-making process*: There must be a clear path in the decision-making process. To use an old Harry S. Truman expression: "The buck stops here." There needs to be one person from the public sector and one from the private sector who is capable and allowed to make decisions. Furthermore, it must be clearly fleshed out which decisions each party will be empowered to make. For instance, it should not be the sole responsibility of the public sector to make decisions regarding private equity matters. Rather, public sector representatives should be allowed to voice their opinions, but final decisions regarding equity matters should be left to representatives of the sector. Conversely, private sector representatives should not be making decisions regarding the permitting process.

6 *Make sure all parties do their homework*: While it should be obvious that it is vital for each side of a P3 to do their homework, it is not always the case. Thus, it is extremely important that this phase is not skipped over, but rather fully engaged. By doing their respective due diligence, each side of the partnership will gain a clear understanding of who the players are; what their motivations are; what the upsides and downsides of a development project are, and what invariable risks loom on the horizon; and, finally, how best to potentially mitigate development risks.

7 *Secure consistent and coordinated leadership*: This follows point 5 regarding the decision-making process, but it is also closely aligned with point 3, that is, knowing the players involved. If a project is long term, say, 5 to 10 years, it is more than reasonable for representatives from the private sector side to expect

leadership from the public side not be assigned to elected officials but to qualified career civil servants who will be able to steer a project through from start to finish. Otherwise, a P3 development project runs the risk of costly delays and/or radical changes that might jeopardize the project.

8 *Communicate early and often*: Communication is an indispensable variable. But not just random communication. A communication strategy must be agreed upon right from the start of a project to leverage public and private support. This involves careful targeting of information to relevant development partners and stakeholders. The strategy must also be inclusive. This is an issue that should be settled at the very beginning of development projects. Some questions related to development communication include the following: Who will be doing the communicating? How will communication be used to improve accountability and development outcomes? And How will communication be used to inform and influence public awareness and support for development projects? This goes back to the above discussion about leadership. Second, how will development communication be structured? How will new media and communications tools be used? Will it be in the form of a simple email, a letter, a conference call, or a site visit? Moreover, how will the various communication devices be used to monitor and measure the impact of communication interventions? Lastly, how often will information be shared, weekly, bi-weekly, once a month, or quarterly? How this communication process will take place is very much tied to the strategic priorities of key players. Regardless of which communication strategies and practices are adopted, it is important to remember that constituency is the key to transparency.

9 *Negotiate a fair deal structure*: The structure of any development deal must be fair to all parties. Again, remember this is a partnership, which means that all boats should rise together. Naturally, the needs, wants, and desires of each partner in a development project will be different. But they should be incorporated into a project that is fair and just for both parties. From both sides, it is important to remember that development projects that are seen to benefit both sides have the potential to catalyze additional projects in the future to both parties' mutual benefit. Thus, it is imperative to ensure that all development projects embrace the principles of equity and fairness. The structure and principles of projects should be codified in writing and signed by all the relevant parties to avoid any preconceived notions regarding the partnership and development outcomes.

10 *Build trust as a core value*: Trust is foundational to the aforementioned principles of successful public–private partnerships. It is akin to an interdependent web that binds all partners in a development project and influences how they cooperate to accomplish their complementary goals. Thus, any development project thrives on the cumulative trust each of its partners invests in it. However, trust must also be verified in the form of contractual agreements, which should be spelt out clearly and early in the partnership process.

Structure of the Book

Addressing the systematic lack of attention to P3 and real estate development in the mainstream real estate literature, this edited book offers an overview of public-private partnerships (P3) and real estate development in the United States. Specifically, it discusses the many and growing benefits of P3 in contemporary US real estate development for both private developers and cities. The book draws on the accumulated knowledge and expertise of academics who study this field, and current practitioners, both in the public and private sectors. To that end, the book fills a void in the real estate literature by presenting P3 development case studies, from which both practitioners (public and private) and academics can learn.

The book is composed of two primary Sections and an Appendix. Section 1 contains five chapters that present an overview of the field of P3. Section 2 also contains five chapters that present case studies by practitioners who were either involved in, or are intimately familiar with, P3 development projects. Following these two Sections is an Appendix. The Appendix includes information on how to make a P3 financially work; an example of a legal document, which corresponds to Chapter 3 (Appendix A); and a mock RFI (Request for Interest proposal), which corresponds to Chapter 10 (Appendix B). Finally, a Glossary of Key P3 Terms lists acronyms/terms that are used in the field of P3.

Section 1

The chapters in Section 1 can best be understood as background chapters. The remaining chapters in Section 1 deal with various development tools that are used, especially Tax Increment Financing, including tools used by the City of Greenville, South Carolina, to change the town from a dead mill town to one the most popular tourist destinations and small towns in the Southern United States; contractual agreements about how development projects are structures; and, lastly, how P3 deals are constructed in terms of their pro-forma.

In Chapter 2, Dr. Jeff Burton discusses the range of public financial tools available to developers in public–private partnerships, focusing on tax increment finance (TIF), business improvement districts (BIDS), brownfields, and eminent domain. Dr. Burton notes that sometimes funding is granted because the public sector desires the construction of a public infrastructure or amenity by a private developer that the market will not finance, such as a contaminant remediation project. Other times, the public sector may want the private sector to behave in a manner that is not conducive to its profit-maximizing logic, for example hiring minority-owned, or veteran-owned, subcontractors.

As discussed, it is important that strong agreements/contracts are established in the beginning of partnerships. This is what Reed Bennett discusses in Chapter 3. Mr. Bennett states that a development agreement and related contracts governing the relationship between public and private actors are key variables in the success of any public–private partnership. The use of contracts within this environment and in connection with public–private partnerships clearly lays out what is expected from each party involved in terms of financial considerations or other assistance or support to be provided, while also addressing a myriad of other important issues such as timing, mandator insurance requirements, and what happens if one or more of the parties in the public–private partnership fails to perform as agreed upon. Some of the commonly known contracts and other agreements of this nature consist of, but are not limited to, the following: Concession Agreements; Offtake Purchase Agreements; Input Supply Agreements; Construction Contracts; and Insurance Arrangements. The chapter serves as an introductory guide to contracts and other written agreements that are used in P3s for the primary purpose of developing real estate.

Nancy Whitworth, the former Economic Development Director of Greenville, SC, discusses the various mechanisms that helped to rebuild the city in Chapter 4. She discusses Greenville's P3 development approach from the effective leadership stage to the financing of the actual redevelopment projects in the city. Some of the financing mechanisms Ms. Whitworth discusses include bonds and districts, including TIFs and BIDs; hospitality funds; parking; New Market Tax Credits; and various other value-added items. Ms. Whitworth does not only discuss these developments in an abstract form; she also provides examples of how the city used them [what is 'them'?].

In Chapter 5, the last chapter in this section, Dr. Burton discusses, in parallel with his discussion in Chapter 2, how to make the numbers in a pro-forma work for a P3 project. In the chapter, he presents a simple method to calculate the costs and benefits of a public–private partnership in a tax increment fixed area/district. The method involves data sets, calculations to understand, and analysis to measure the complementary benefits to developers, municipal governments, and local communities.

Section 2

While Section 1 primarily discussed the background to P3, Section 2 discusses how those background ideas and thoughts play out in practice. The cases discussed in this section include the development of a minor league baseball stadium; innovation districts; a P3 development of an industrial complex; a small-scale development that tapped into CRA funds to develop a project; and lastly, the US Navy partnering with local municipalities to develop underused base real estate property.

Beginning this section, in Chapter 6, James Frazier, a commercial real estate broker with RealOp Investments, and Dr. Stephen Buckman discuss the partnership that helped to build a small minor league baseball stadium. Echoing Ms. Whitworth in the previous section, the chapter examines the building of Fluor Field (a minor league baseball stadium) as a catalyst for the redevelopment of the West End of Greenville, SC. The stadium helped to spur real estate development in the West End, which in turn helped to create the third leg of the development of Greenville's downtown: Hyatt-Falls-West End. The chapter shows how, without P3, development could probably not have been accomplished in Greenville. It also shows how other communities, especially smaller communities (Greenville has a population of 70,000 people) can follow Greenville's blueprint for minor league stadium development.

Chapter 7 follows in the discussion of TIF and BIDs in Chapter 2 by discussing the concept and practice of innovation districts. Dr. Carla Maria Kayanan of University College Dublin and Dr. Patrick Cooper-McCann of Wayne State University show that, especially since the 1970s, numerous state legislatures in the United States have authorized place-based economic development tools, including tax increment financing districts, business improvement districts, and enterprise zones, that enable local governments to partner with developers and business owners to spur economic growth within designated areas. Increasingly, local leaders are using these popular but controversial tools to advance a new economic development concept, the innovation district, which promises to generate next-economy jobs in dense, mixed-use, live–work–play settings. The chapter explains the innovation district concept and then critically examines its implementation in four locations in the United States. It concludes with a discussion of the concept's promise and limitations relative to prior approaches.

In Chapter 8, Brockton Hall, Senior Vice President of Colliers International Industrial division, presents a case study of a site selection, entitlement, and acquisition process for a large-scale industrial development which took place in the southeast United States from 2017 to 2019. During this process, the developer worked intently with the local economic development authorities and government officials in the area of the subject property to develop a public–private partnership strategy which led to the development of over 2 million square feet of industrial buildings that were mutually beneficial to both the developer and the municipalities where the project took place.

Chapter 9 discusses working with Community Redevelopment Agencies (CRAs) to get a development done. In this chapter Joe Bonora discusses how a small-scale developer can tap into CRAs to obtain incentives that will make a deal work, which otherwise would not work if there were no help from the public sector. He highlights two key ways this is done: 1. By building strong interpersonal relationships and 2. By doing your homework and understanding not how the CRA can help you but how you can help the CRA. He discusses these factors through the description of three case studies of developments he was involved in that partnered with the local CRA.

Lastly, in Chapter 10, Lieutenant Commander Mike Yeary of the US Navy and Dr. Stephen T. Buckman discuss how the Navy could partner with nearby communities to find the best ways to develop their underused real estate. The chapter discuss how Naval Facilities Command (NAVFAC) requires improved means and methods for real estate and planning professionals to identify underutilized installation real estate property and improved policy on how best to implement projects on this real estate. Within this context, U.S. installations have a long history of collaborating with surrounding communities through a variety of partnership agreements with local, state, and civilian players. These public-to-public (PuP) and public–private partnerships (P3) provide a wide range of benefits, including reducing costs, increasing services, and sharing capabilities to accomplish shared goals. With recently declining defense budgets, installations need to find novel ways to sustain military facilities, provide essential services, and provide for the full host of military, civilian, and contractors who inhabit bases and enable missions. By leveraging the development community via P3 best practices, combined with identified underutilized real estate opportunities, the Navy can develop projects that better serve military installations and surrounding communities.

2 P3 Tools

TIF, BIDs, Brownfields, and Eminent Domain

Jeff Burton

Introduction

Sometimes the community needs private investment in higher risk areas that market forces will not finance naturally. A public capital infusion into the financial equation may be that puzzle piece that completes the development capital stack. Brownfield site remediation and its redevelopment or low to moderate housing accommodation are two such examples. Other times, the public sector offers incentives to entice the private side to improve their projects physically, when not required by standard codes. One example could be the developmental addition of Low-Impact Development (LID) in an older urban area with no pervious surface, offering stormwater pretreatment capacity where none exists, and is not required, due to credits given for grandfathered impervious surfaces. A second example might be a development's upgraded exterior to meet certain aesthetic overlay requirements, like a historic district design code. Yet, other times, the public sector might want the private sector to behave in a certain manner not conducive to their profit-maximizing proformas, for example, the preferential hiring of minority-owned, or veteran-owned, subcontractors that may not be the development's lowest bidder. Following are some practical examples of each.

For instance, the Mayor of Palmetto, Florida supports the idea that all multifamily midrise apartments built in the city's flood and high wind area should be constructed with concrete masonry units versus wood. She asked for an incentive policy to be developed. To her, this type of masonry construction was deemed more resilient to wind, flood, and storm surge scouring than wood. At the time, there was additional cost for this added effort. The cost was addressed through an increment revenue incentive. In the event of a storm event, that added improvement may save far more than it cost, and may reduce the developer's insurance premium, or save lives. It is worth it!

On the other hand, the City of Fort Myers, Florida wishes to require developers, building in their "blighted areas," to hire 15 percent of their subcontractors as minority-, woman-, or veteran-owned (MBE, WBE, or VBE) companies. National retail chains typically have a "cookie cutter approach" to their stores that uses the same general contractor and subcontractors to regulate their cost through design and building familiarity and efficiencies and economies of scale estimate requests.

DOI: 10.1201/9781003222934-3

That incentive for that additional 15 percent guaranteed change was set in place almost a decade ago and continues to benefit the developer and the economically distressed neighborhoods.

Addressing multiple brownfield sites, the State of Florida offers up to 75 percent of the remediation cost returned to the developer as state tax credits, which can be sold one time, for cash. In Tampa, Florida, those tax credits incentivized the redevelopment of an old dilapidated industrial complex into the State's first IKEA. These are real public–private partnership (P3) examples that you may experience and take advantage of.

In writing this chapter, some assumptions were made in framing this discussion scope. Other types of tax sources may be used in the practicing world, but it is assumed that for this chapter the tax revenue source discussed is ad valorem property taxes. Also, that the higher the governmental hierarchy (local, state, and federal) from which the funds are sought, the more and varied strings are attached. Finally, over my years of professional experience and continuing education, it has become very clear that our society's current profit-at-all-costs motivated development methodology is unsustainable. A new, more evolved "sustainable (re)development" approach to development is in order and needed. This new approach is holistic and addresses other valuable concepts such as the environment and society. I did not always think this way.

As a young Palmetto, Florida City Councilman, in the early 1990s, I supported economic development in my hometown at any cost, I was ignorant and blind to the anthropogenic effects that mindset causes. Now I am older (so much older), and hopefully a wiser professional and academic. My doctoral research used a sustainable development theoretical framework to measure tax increment-financed (TIF) brownfield remediation and redevelop state policies across the United States. This work focused me and now guides my professional and personal thoughts and actions. This theoretical framework, developed in 1987 by Dr. Edward B. Barbier, introduced a Venn diagram with economic, social, and environmental ingredients that form the sustainable economic development recipe (Figure 2.1). Our past and current development models skew heavily to an economic gravitation that is genetical ingrained into our U.S. society, but at a massive expense to our environment and society. How does this sustainable recipe frame your development?

In 2021, collecting 59 state TIF statutes and using a mixed-methods analysis, I framed them with the sustainable development theory (Figure 2.2). The same scientific methodology was used regarding 47 brownfields statutes, which is discussed later in this chapter (Figure 2.7). Figure 2.2 is the proportional and dynamic Venn chart for the TIF law sustainable development elements. The chart displays the overarching strength of the economics shadow at the cost of the other elements. Notice the TIF law's social element size compared with that of the others. TIF laws are predominantly eco-environmental.

How can these models change for a more balanced and sustainable (re)development? How would the balance of a housing sustainable development Venn chart look? What other development elements could be framed using this theory? Finally, consider layering Venn models, so that TIF finances brownfield remediation for

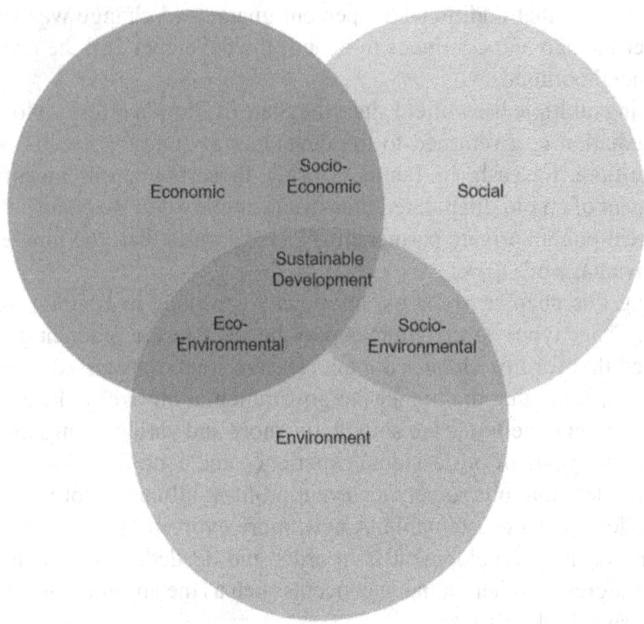

Figure 2.1 1987 Barbier sustainable economic development Venn chart amended By Jeff Burton.

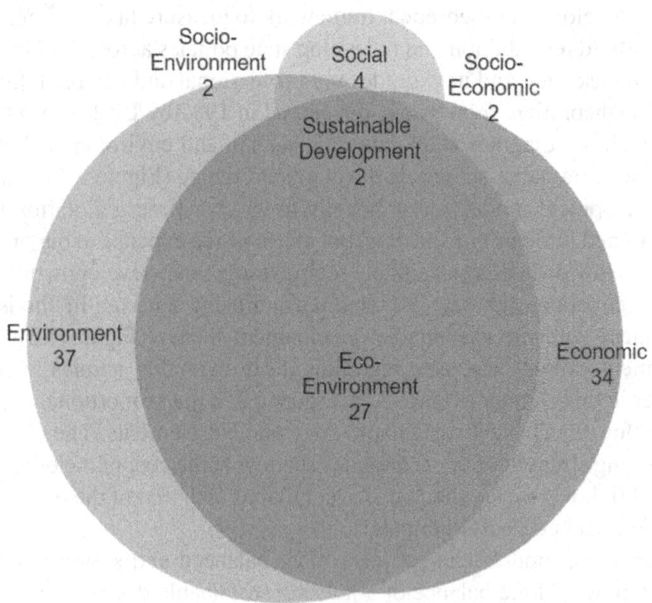

Figure 2.2 Proportional and dynamic state TIF statute sustainable development Venn chart.

houses that are resilient to high winds, floods, and fire! Would the sustainable outcomes be cumulative or multiplicative? Think of the real estate value that kind of home would have in hurricane-prone Florida. What type of property and casualty insurance premium would be offered? Sustainable (re)development can lead to more profitability!

Tax Increment Financing

Tax increment finance (TIF) is a tool originally created by local elected officials and state legislators to match federal housing and urban renewal dollars without leveraging existing or raising new taxes and upsetting voters.

The 1949 U.S. Housing Act changed the landscape of the Federal government's financial involvement in local housing and economic development policy. Prior to that law and starting in 1936, the Fed focused on housing, responding to the Great Depression. These new policies attempted to mitigate lost mortgages from "sunken" middle-class, predominately white Americans, while continuing to ignore the housing needs of the lowest of the low incomes, primarily African Americans. The 1949 law expanded the Federal scope to include job creation for millions of returning World War II veterans. These returning GIs were predominantly white, and the policies excluded many minority citizens from the benefits, while at the same time taking their communities, bulldozing them, and reselling them at discount to developers (urban renewal). Starting in the 1940s and continuing through the 1970s, tax increment was used to "buy down" developed, but usually slum or blight-stigmatized communities. In the end, many of these "buy down" properties ended up vacant from a lack of private investment redevelopments. How did tax increment finance support these policies?

Although legislatively approved in Wisconsin in 1942, just prior to the 1949 U.S. Housing Act, tax increment finance was first used in California in 1952. By 1970, only Minnesota, Nevada, Ohio, Oregon, Washington, and Wyoming had added tax increment legislation. Richard Nixon ended urban renewal in 1973. The "Reagan–Thatcher" era of the 1980s shifted economic development funding from the Federal level to the state level. Tax increment use expanded to 49 state adoptions and is now the most prevalently used public–private partnership funding mechanism.

Tax increment policy is predominantly created at the state level of government and takes on many policy forms. These may vary by funding source or type, by creation term, and expenditures. Ad valorem, or local property taxes, are the primary tax increment funding source. This creates a direct relationship between the increased value of the development to the increased value of the increment. Though not as common, sales taxes may be used to create the increment. This source may have an indirect relationship with the private investment, or none. Though we will discuss different sources, this chapter will focus on property tax, or ad valorem-based TIF.

Depending on the state, some policies focus on a specific project, while other laws determine an area. The first type tends to focus on economic development, while the latter may be the answer to slum or blight area issues. Regardless of the geographic area, terms usually extend into decades. Locally, how does this work?

Local taxing authorities are required participants in the tax increment methodology. Depending on the TIF's source, it may be a city, county, special district or authority, school district, hospital district, or fire district. Measuring the estimated taxes for a property requires the knowledge of its millage rate. This rate is usually multiplied by a property's value divided by a number of years divisor. For example, a property worth $1,000,000 in a city that has a 4.5 millage rate per every $1,000 of the property's value will pay $4,500 in property taxes that year.

Property Tax = 4.5*(1,000,000/1,000) = $4,500

The property value is determined by a property appraiser. Most states have their own methodologies for determining this value. The "real property" value includes the land and building on the land, not to confused with "tangible personal property," which are investments made on the land that can be sold and removed. Also, the two prominent real property value methodologies are actual value and income value.

1. Actual value is determined by the present value of the land and facilities. This value is determined through comparable sales of similar properties in the area.
2. Income valuation is determined from buildings that derive rents and leases. These may be multifamily rentals, office complexes, and retail strip malls.

Some states may offer property tax exemptions based on a specific criterion. These exemptions must be taken into consideration when calculating TIF values.

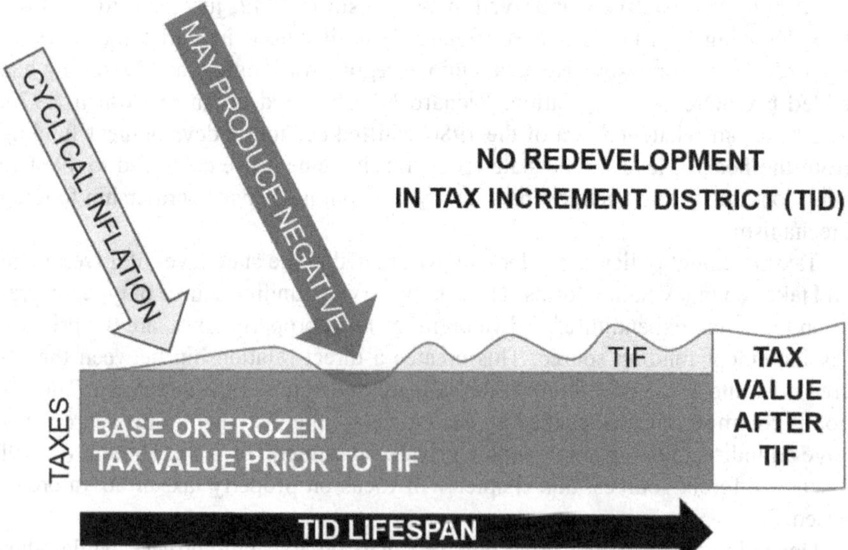

Figure 2.3 TIF growth with inflation only.

Exemptions can significantly affect the property values. Churches tend to have a complete property exemption for their religious buildings, rendering their investment mute when it comes to TIF. Some states do not exempt church-owned apartments, or other buildings that produce an income. Some states exempt county, state, and federally owned lands, even if leased and producing an income. How are these properties evaluated for TIF? It starts with the base year.

The base or "frozen" year is the year in which the TIF was formed. Any taxable value prior to the creation continues to flow to the taxing authorities. Any incremental value created after this year is given to the redevelopment authority. Increment grows through general inflation or private redevelopment investment.

The redevelopment defines the evolution from what was on the property to what is. Following are examples of varying land use evolutions and their taxable results:

1. No TIF

    1.1 Totally exempt to totally exempt equals no TIF.

    If the government owns the site and redevelops it into something new, and with a higher property value, the site still pays no taxes. This estimate can also be used for church- and non-profit-owned properties.

    1.2 Non-exempt to exempt equals no TIF.

    If the government buys a privately owned property and redevelops it to a public use. Also, a church or non-profit buys privately owned land and redevelops it to their primary use. The land is exempt from property taxes.

    1.3 County, State, and Federal to leased non-exempt equals no TIF.

    In most states, federal, state and county land is non-taxable no matter the use. This includes the lease of the land to a private for-profit use. City land does not usually have the same exemption.

2. Inflationary TIF

    2.1 Non-exempt to non-exempt equals inflationary TIF.

    Land values, on average, increase over time. A privately owned property will increase in value due to inflation. This method applies to residential, commercial, and industrial uses. Some states authorize homestead exemptions and/or a maximum annual percentage growth rate (Florida's Save Our Homes law).

3. Redevelopment TIF

    3.1 Totally exempt to non-exempt equals TIF.

    If a church property, which existed when the TID [Tax Increment District] was created, sold for a private sector redevelopment in a downtown, the new site would create a maximum TIF. The church site's exemption provided a base value of $0, the TID would receive the maximum TIF.

3.2 Non-county, state, and federal to leased non-exempt equals TIF.

If the city leased land next to the county convention center, to a hotel, the redevelopment would general maximum TIF, as the base year would be $0.

What factors affect property values? Ad valorem tax increment is created by inflation/deflation and investment/disinvestment. Inflation growth is cyclical. The economy expands upward to a peak and recesses downward to a trough. When private real estate investment occurs, the potential for the property's value to rise is significant. So how are taxes calculated?

Remembering our property tax calculation:

Property tax = milage rate × (property value/divisor)

A tax increment finance calculation might look something like this:

TIF = Property Tax (TODAY) − (Property Tax (BASE YEAR))

TIF is a powerful development finance tool. It should not be used lightly, and when used, remember that the web-slinger's rule that with great power comes great responsibility applies. Without the private developer's investment, there is no tax increment. Theoretically, the public–private partnership "self-funds" itself, but in many cases, the local government does not really understand this financing, how to

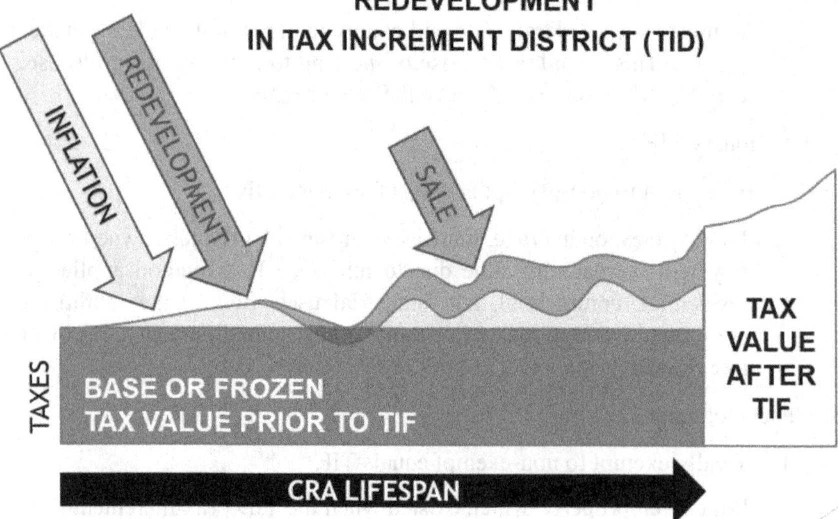

Figure 2.4 TIF growth with redevelopment.

calculate it over time, or how to mitigate for external economic shifts. This lack of understanding can lead to some serious financial problems. See Chapter 5 of this book for a cost/benefit calculation.

The increment spent within the Tax Increment District (TID) is created within the district, and does not take taxes away from the rest of the city. Many states require TIF to be used in "slum" and "blighted" areas, to lift the communities up to city standards and reduce general municipal spending on services such as police, fire, and emergency medical services.

A mentor of mine once described me as a bright lightbulb when I entered a room. I thought that was great, until he said that if I ever focused my light on one spot, I could burn through anything like a laser! Then I thought that maybe being a lightbulb was not such a great compliment. TIF can be a very bright light. When used strategically, it is even a brighter, powerful laser. It can address economic development inadequacies, housing, environmental, and cultural issues. In some states it can even cross the public/private line. The valid argument to this positive TIF attribute is that it is rarely used strategically or sustainably. In many cases TIF is no more than a short-term political slush fund.

TIF may be considered by some, especially those living in the jurisdiction but outside of the TIF area, as taxation without representation. The tax increment withheld for the TID may require that services throughout the governing district cost more to the non-TID customers. These taxpayers may argue for their loss of representation regarding the taxes not paid into the general fund. TIF advocates argue that the negative issues being addressed by the TIF dollars came from years of fiscal neglect by the overall jurisdiction in the first place.

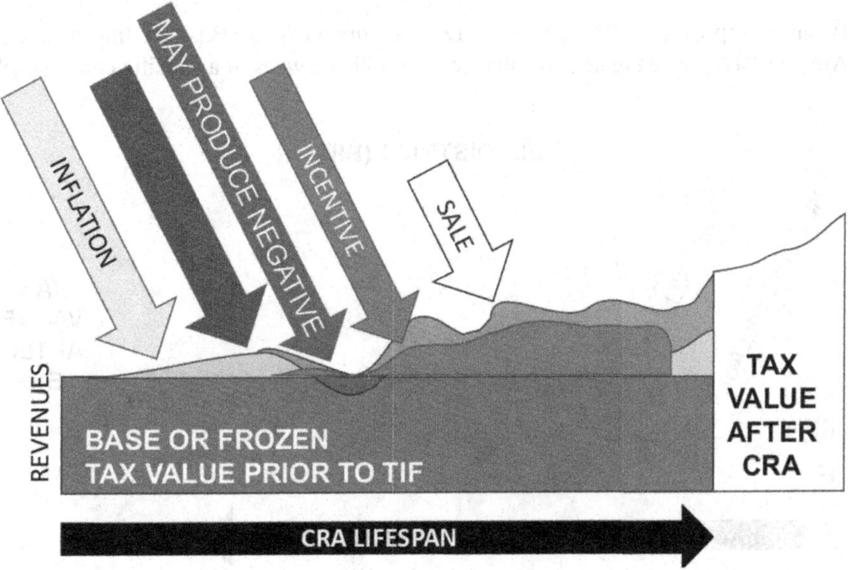

Figure 2.5 TIF model with redevelopment and incentive.

In some cases, TIDs have been created at the height of the market, just prior to recession. This timing may cause the TIF to go negative, meaning that the market-high taxable values at creation are now deflated, and the TIF is less than zero dollars. Depending on the severity of the deflation, the TID may require a restart.

Another potential TIF negative is that, during its creation, the geographic boundaries may not include the right mix of land uses. Some states, like Florida, may have placed an artificial taxable value ceiling on homeowner occupied residences (3 percent annum) or commercial properties (10 percent annum). A TID with just these types of land uses may never grow into a mature authority. Are there other forms of TIF?

Most states program the financing tool for public projects in support of private investment. These types of projects include roads, sewers, water, stormwater, fire or police services, or Internet connectivity. A smaller percentage of states reach across public/private sector lines and provide direct incentives to the investment.

Most states use tax increment financing to entice (re)development through the strategic use of localized taxable real property improvements through public expenditures around those investments. In the State of Florida, due to some unintentional legal circumstances, increment revenue, as it is termed, was deemed to be not a tax by a 1980 state supreme court legal decision (*Miami Beach v. State of Florida*) and a reaffirmation in a 2007 case (*Strand v. Escambia County*) by the same court. Not being a tax, the revenue is not required to meet the state constitution's public funds for public uses litmus test and can be spent directly with the private sector. Therefore, the funds are not tax increment finance or TIF but are increment revenue.

Business Improvement Districts

Business Improvement Districts, or BIDs (sometimes called a Business Improvement Area, or BIA) were created to enhance the public services in a specific commercial

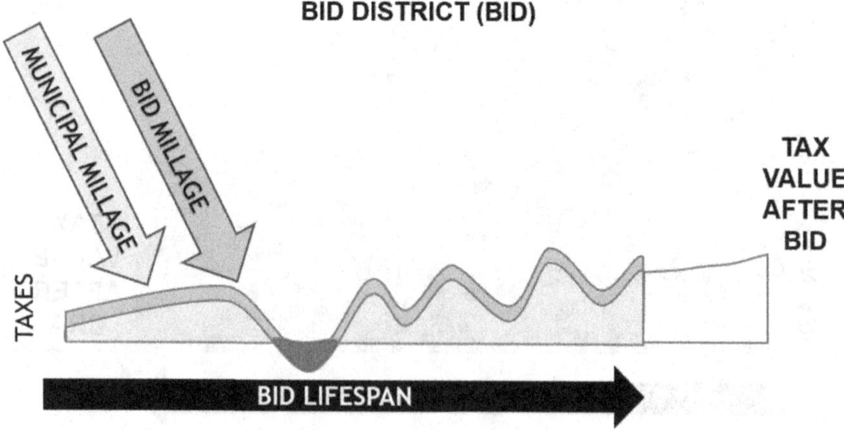

Figure 2.6 BID revenue over time.

area, usually a downtown. Services are usually above and beyond local public levels and may include safety, maintenance, and events. Business owners may also want greater control of their area's future public investments. These owners also may want a business plan that impacts their properties. By uniting as a group, the businesses may also have one stronger political voice. How did BIDS start?

BIDs were first created in Canada in 1970, while the first U.S. BID was created in 1974 in New Orleans. In most cases, the BID is managed by a non-profit, with the paying business owners as the shareholders. I see BIDs as the evolutionary successors to slum and blight TIDs. TIDs were created to address social problems like disease, poverty, and neglect. This was the mantra of the post-World War II decades, especially with the Federal Government picking up most of the bill. BIDs are economic in nature, requiring little to no federal financial help. BIDs are more concerned with the economics of its area. How do BIDS compare to TIF?

1. BIDs offer similar types of benefits to tax increment financing. The main difference with a BID is its tax source. Where TIF is usually generated by growth in a property's value, BID income depends on the increase in the millage rate. BIDs are usually allowed to add one to two mills above the current tax rate imposed by the local government. Where TIF is "self-funding," BIDs are "self-taxing."
2. In the same framework as TIDs, BIDs usually require a spending plan.
3. Like TIF, BIDs generally are used to preserve property values and create stronger relations with the area's business owners. Bids also can provide promotion and marketing. Both entities can usually monetize their incomes, depending on state law.
4. Like TIF districts, BIDs have been scrutinized for their lack of assessment and accountability. The spending of local government dollars through a one-off non-profit may lose focus and increase mission drift. Also, successful BIDs may lead to rent escalations, which may promote gentrification.
5. Where TIDs were founded on affordable housing, BIDs have been criticized for policing the homeless from their economic centers.
6. Since a BID may require the levy of additional property taxes, local elected officials are not very fond of them. Since a BID requires an increase in taxes, a referendum of its area is usually required along with a plan of the use of additional taxes.

As was noted above, BIDS are self-taxing, charging each property with an "above and beyond" self-imposed property tax. The property tax scheme may look like this:

Property tax = (BID Millage) + (local government millage) × (property value/divisor)

BIDs do have value in the public–private partnership model. First, they can carry out studies, plans, and reports that may indirectly assist the developer with one-off public improvements, benefiting the private investment. The BID may also fund and maintain, above and beyond the local government level of service, public realm improvements surrounding the new investment. These non-profit entities may also invest in historic preservation and market business in their area.

Some BIDs provide for an exemption for their added tax as an incentive for new businesses. BIDs have also used their resources to provide façade grants for existing businesses to improve their aesthetic appurtenance.

Brownfield Remediation and Redevelopment

A brownfield site is a parcel of land, the reuse, expansion, or redevelopment of which may be complicated by the existence of, or the perception of, hazardous substances, contaminants, or pollutants. This chapter section discusses these properties and how they can be encompassed in a public–private partnership. Why is a brownfield program needed?

Prior to the creation of a U.S. brownfield law (2002, The Brownfields Revitalization and Environmental Restoration Act of 2001), the country dealt with numerous environmental issues. Due, in part, to the 1977 Bridgeport, New Jersey chemical waste treatment fire, the 1978 Niagara Falls, New York, Love Canal neighborhood contamination, the 1978 Valley of Drums, Kentucky investigation, and the 1983 Times Beach, Missouri dioxin contamination and subsequent city disincorporation, the U.S. Congress adopted, and President Jimmy Carter signed, the 1980 Comprehensive Environmental Response, Compensation Liability Act, or CERCLA.

CERCLA gave the United States Environmental Protection Agency, or USEPA, the ability to create standards by which to measure allowable and harmful exposure to different pollution for different land uses. The law also gave the USEPA the legal and accountable apparatuses to enforce those measures.

The law included a superfund by which the USEPA remediated highly polluted sites and then sued the owners and other ancillary persons of potential liability. These activities caused a rash of private property abandonments, especially in urban municipal areas. In the end, the local governments involuntarily took ownership and the stigmatized properties sat.

From the negative public reactions to these highly regulatory actions, lawmakers amended to CERCLA a more voluntary remediation program in 2002 called the Brownfields Revitalization and Environmental and Restoration Act of 2001. This Act provided economic redevelopment opportunities for non-responsible property owners to clean the land for economic benefit.

A key element of the program is the federal/state government partnership. This relationship led to many of the states adopting their own statutes with variations between them.

The USEPA provides assessment and remediation grants, as well as revolving loan fund programs available for sites and for local governments.

Many of the states use some form of corporate or real property tax credit incentive system. There is typically some type of liability risk forgiveness for new owners and financial partners. States may also set aside grants for the actual property remediation. How does a brownfield remediation work?

Most state laws do not allow the existing owner of commercial property use of brownfields programs unless an environmental survey was completed at their purchase closing. Most financial entities will not offer loans without one also. Brownfields remediation starts with the knowledge or assumption that a property is polluted or may be polluted. This stigma usually comes from the property's land use history. The site may have been a gas station, dry cleaner, or other known polluter. An Environmental One survey provides a paper trail of the land's prior uses. If a potential source of pollution is discovered, an Environmental Two is almost always required. In rare cases, when the pollutant is naturally occurring, the state may exempt the property from further remediation. If the property passes the first survey, it moves to closing. A second survey usually requires onsite physical sampling of soil and water, below the water table.

Remember the TIF law sustainable development Venn chart, Figure 2.7, displays the same outcome for brownfields statutes. Choosing from economic, environment,

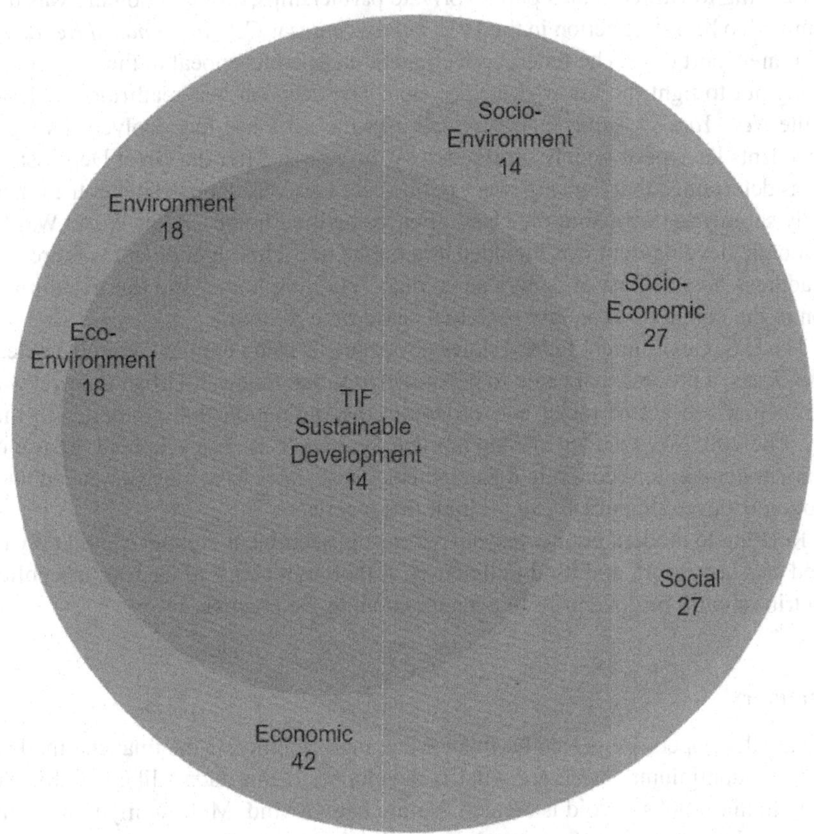

Figure 2.7 Proportional and dynamic state brownfields statute sustainable development Venn chart.

or society, which Venn element would be least represented in U.S. brownfields statutes? Environmental codes were found in only 18 of the 50 American state laws. Twenty-seven states identified some type of social codes and almost all of the brownfields statutes were economically driven.

Eminent Domain

Eminent domain, taken from the Latin term *dominium eminens* (supreme ownership) gives the government the right to take and acquire property for public use. The term was first introduced in 1625 by Dutch jurist Hugo Grotius. Eminent domain is not condemnation. Condemnation is the government act of demolishing a structure due to a state of unsafe or unlivable conditions, but the land title remains with its owner. United States founding father James Madison argued that proper compensation should be paid for property taken through eminent domain. This notion appears in amendment 5 of the U.S. Constitution.

Relating to United States public–private partnerships, eminent domain was determined to be a state action in the 1935 *United States v. Certain Lands of Kentucky* Supreme Court case. The federal government dropped its appeal to the high court, opting not to fight the lower court decision. The decision was reaffirmed in 1936 in the *New York Housing v. Muller* case. As the act's use has evolved, its legal precedents have been widely scrutinized by the courts. After the Great Depression, it was determined that housing was a public use. This was done to shelter the financially submerged white middle class, after losing their homes. After World War II, economic development was included as a public use. This amendment was created to address the millions of American soldiers returning home, and the transformation of the country's economy back to a peace time rhythm.

The U.S. Government funded states governments using eminent domain to clear slum areas, a process that came to be known as urban renewal. Urban renewal was the legitimized act of racial and economic discrimination that continues to this day. The 2005 *Kelo v. City of New London Supreme Court* case upheld the use of eminent domain for economic development. Most states have since amended their economic development laws to prohibit this practice.

Relating to modern public–private partnerships, eminent domain should only be used as a last resort, and used judiciously. A thorough check of each state's policy doctrine should be done prior to even mentioning the practice.

Summary

Private developers are driven by finance and marketability. If the financial markets fund condominiums, those are what is developed. If that fades, like it did in the early to mid-2000s, a void is created. Nature hates a void. Multifamily rental units become the next hot market; developers build those now. Public redevelopers, especially those fueled by TIF, want that new development in their old areas. There is a joke that notes: the kid was so ugly, the parents put a porkchop around the kid's neck to get the dog to play with him. In an ironic sense, that is a government-side redevelopment economic P3.

How do we know if the deal is good for all parties? There must be a cost/benefit measurement used today that can shed light on this question. You will find it in Chapter 5.

3 Contracts

The Heart of Public–Private Partnerships

Reed L. Bennett

Introduction

While sometimes an overlooked aspect of any transaction involving the public and private sectors, contracts are one of, if not the most, important components of any successful public–private partnership, especially those which involve the development of real property due to the myriad of risks associated with an undertaking of this kind such as significant costs and lengthy project timelines. Additionally, as with most other more well-known types of public–private partnerships (or "P3s") such as Concession Agreements or Offtake Purchase Agreements that are common with newer infrastructure projects, these relationships are also subject to concerns related to changing political headwinds, the potential failure of one or more parties—either the public and/or private participant—involved in a particular P3 project or transaction to perform their respective agreed-upon responsibilities, and the occurrence of unforeseen and/or unexpected events such as a hurricane, tornado, act of terrorism or war, or global pandemic—such as the COVID-19 pandemic which has caused significant and widespread issues throughout the world since it was first discovered in 2019.[1] Just look at Amazon's previously announced plan to build another headquarters in the Long Island neighborhood of New York City as a recent example of how powerful political headwinds can be and how quickly support for a significant proposed real estate development project can disappear, along with the potentially lucrative financial incentives and other benefits that were previously promised as well. Or more recently, in South Carolina, the alleged failure of the City of Rock Hill to timely fund public infrastructure improvements that were previously agreed upon in exchange for the Carolina Panthers constructing their new $800 million headquarters within the city limits of Rock Hill, SC, which has resulted in the NFL Franchise's special purpose entity behind the development of this proposed project terminating the various agreements it was a party to with the City of Rock Hill (Benjamin, 2022).[2] Since then, GT Real Estate Holdings, LLC, the special purpose entity affiliated with the Carolina Panthers and its owner David Tepper, has stopped work on the project just as vertical construction was getting underway and has filed for Chapter 11 bankruptcy protection, with the dispute between the City of Rock Hill, SC and GT Real Estate Holdings, LLC is playing out in court, with many years of litigation and legal fees likely ahead for both parties (Brierton et al., 2022).

DOI: 10.1201/9781003222934-4

Ultimately, when it comes to public–private partnerships primarily focused real estate development projects, written contracts and other similar agreements are crucial to the success of such projects for both the private and public parties that are involved, as they memorialize the respective roles, responsibilities, and obligations of each party entering into the public–private partnership or transaction, while also addressing what happens in the future if either, or even both, parties are unable to perform or complete their respective responsibilities under the written agreement governing their relationship. Although written contracts can be time-consuming and painstaking to negotiate because of the various parties involved in a public–private partnership and just how much the interests each of these parties may have could differ, the benefits of a thorough and well-drafted written agreement greatly outweigh any preconceived cons, and simply cannot be understated enough.

After providing an overview of some of the basic contract law principles that are relevant to public–private partnerships, specifically those primarily involving real estate development in the United States, and discussing some of the most common types of agreements and contracts governing public–private partnerships, the remainder of this chapter focuses on several key contractual provisions to consider incorporating into any future written agreements or contracts governing a public–private partnership created for the principal purpose of developing real estate, and other legal issues worth critically thinking about in greater depth before formally entering into any public–private partnership.[3]

Contract Law Basics with Respect to Public–Private Partnerships

While there is no widely accepted definition as to what constitutes a valid and enforceable contract among legal scholars—they all vary slightly, just like the laws in most states regarding the same—at its most basic level, a valid and enforceable contract between parties is comprised of the following three essential elements: (1) offer, (2) consideration, and (3) acceptance. One of the foremost authorities on legal definitions, *Black's Law Dictionary*, also defines a valid contract in similar terms stating that it is a "[a]n agreement, upon sufficient consideration, to do or not do a particular thing" ("*What Is Contract?*," n.d.). It is important to note though that a contract does not have to be in writing to be considered valid and enforceable by a court of law under certain circumstances; however, if you only take one thing from this particular chapter, it is this—make sure to memorialize, or document in writing, any agreement between a public municipality or other governmental agency and a private developer in a detailed written contract that helps eliminate any room for ambiguity or disagreements later on down the line. Additionally, a detailed written agreement among the various parties involved in a particular public–private partnership also helps ensure that the legal concept known as the statute of frauds is satisfied, which is a general requirement that certain contracts must be in writing and executed by all parties to the respective agreement. Specifically, since the statute of frauds is generally required when an agreement between respective parties involves the sale or transfer of real property and leases longer than a year in length, either or both of which are likely to be present in an agreement relating to a real estate development project, a written agreement or contract between the public and private parties entering into any public–private partnership is *strongly* encouraged (emphasis added). Lastly, a detailed written agreement should also help mitigate the risk of a disagreement such as the significant one currently heading to litigation between the NFL's Carolina Panthers and the City of Rock Hill, SC; however, unfortunately, even a thoroughly negotiated and well-drafted written contract still cannot completely eliminate the possibility of a dispute such as this arising.

Another thing to be aware of with public–private partnerships and the contracts governing the relationship between the public and private parties involved in the same real estate development projects or transactions—preferably agreements that are in writing—is importance of the individuals signing on behalf of both the private and public parties having the requisite authority to act on and bind the respective party each represents. Usually, this is a bigger concern with the representative of the public entity that is a party to the written agreement governing the public–private partnership lacking the necessary authority to act on, bind, and/or sign on behalf of the local municipality or other governmental agency he or she has been elected to represent or hired to work on on its behalf; however, depending on the jurisdiction and the organizational documents governing the particular municipality or governmental entity in question, usually this issue can be resolved by having the local municipality or government agency publicly adopt an authorizing resolution which explicitly approves this public sector entity's involvement in the proposed real estate development project. In addition to broadly approving the proposed project and the public body adopting the authorizing resolution involvement in it specifically, typically a resolution such as this will also expressly ratify the agreement between the governmental entity and the private party relating to this particular project or transaction (the "P3 Contract"), and clearly designate a certain public official—preferably two if possible—who is authorized to execute and/or witness the P3 Contract, as applicable, and any other agreements, certificates, instruments, and/or documents as are necessary to consummate the transaction contemplated in the authorizing resolution, and to take any other such actions which may be required in connection with this particular proposed project or transaction.

It is also important to mention here that the specific name of a particular written agreement governing a public–private partnership isn't important, but rather the content of the written contract itself. This is important to note because there are a number of different written contracts that have similar names yet include drastically different contractual provisions, which is especially true in the case of public–private partnerships focused on real estate development. For example, in the State of South Carolina, contracts known as "Development Agreements" and "Fee-in-Lieu of Tax Agreement" (or "FILOT Agreement") may seem quite different based on their respective names; however, these two agreements have substantially the same purpose of trying to encourage greater capital investment in a certain area or within the jurisdictional boundaries of a particular municipality through the provision of certain incentives from a local governmental agency or other public body. Further, the same idea also holds true when trying to compare two different states, as the State of Georgia doesn't have FILOT Agreements, but rather has a contract which includes most, if not almost all, of the same provisions known as "Payment in Lieu of Taxes Agreement" (or "PILOT Agreement"), which a substantial number of states also utilize in a somewhat similar manner subject to their respective laws and regulations, such as the State of Tennessee for example.

Finally, in a similar vein, it is important to note that sometimes multiple different contracts or written agreements might be required to address all the different aspects of a particular public–private partnership established primarily for the purposes of the development of real property, while in other situations only a single written agreement might be needed in order to comprehensively address the entire relationship between the public and private parties involved. For example, with the recently completed mixed-use development commonly known as Camperdown in Greenville, South Carolina, this particular public–private partnership resulted in at least two different written agreements between the project's master developer, Centennial American Properties, and the City of Greenville, SC. Specifically in the case of Camperdown, there was a development agreement which addressed most of the proposed development project that was substantially completed in 2020, such as the establishment of a synthetic tax increment-financing district around the entire project site and the requirement that certain public improvements be constructed and where, as well as a separate reimbursement agreement that dealt with primarily certain utility upgrades and water quality improvements that the City of Greenville, SC wanted done on site, and providing compensation to the master developer of Camperdown in exchange for undertaking these desired infrastructure upgrades and improvements during construction. There is no concrete rule as to how many written agreements might be necessary to govern the many different aspects of a relationship between public and private sector parties—let alone possibly multiple of each, if not more—relating to the development of real property, and this decision will ultimately be made by the respective parties' legal advisors who are responsible for negotiating and structuring the public–private partnership.

Common Types of Agreements Governing Public–Private Partnerships

Even though the term "public–private partnership" has been researched and discussed extensively for many years, there is no widely accepted definition of it since its meaning varies greatly depending on the context in which it is used, which might help explain why the latest version of the *Merriam-Webster* dictionary doesn't even include it.[4] Nevertheless, since the term public–private partnership is most often used in connection with infrastructure projects such as toll roads and privately constructed power stations that sell the electricity they produce to regulated power utilities, when other authorities have defined what exactly this term means they have usually mentioned infrastructure explicitly, or have been similar to *Encyclopedia Britannica*'s definition of the term which stated the meaning of a public–private partnerships is a "partnership between an agency of the government and the private sector in the delivery of goods or services to the public." While the term public–private partnership has been utilized in the United States to refer to many other different types of relationships between public and private parties such as the public sector's funding of private-sector research activities, due to the significant role public–private partnerships play in the infrastructure sector, most of the well-known types of written agreements and contracts in this space today relate to either transportation or the provision of an essential public good or service such as potable water or sanitation services. Specifically, some of the more well-known types of agreements and contracts governing public–private partnerships today include, but are not limited to, Concession Agreements, Offtake Purchase Agreements, Input Supply Agreements, Construction Contracts, and Operation and/or Maintenance Agreements, all of which are discussed in greater detail below.

First, with respect to concession agreements, these are contracts in which a private party—most likely a special purpose entity created solely for a particular project—designs, builds, finances, and operates a new public infrastructure project or facility for a governmental entity or agency, in exchange for the private party, or one of its affiliates such as its parent organization, being allowed to levy a user charge or fee for a finite period of time—also known as the "concession period"—on the members of the general public who use this particular project or facility during the concession period (Yescombe, 2013). An actual example of this type of agreement would be Georgia's State Road and Tollway Authority contracting with a private party to build and maintain new toll roads around metropolitan Atlanta in exchange for the private company being able to keep the revenue generated by the tolls for a set, predetermined number of years (Wickert, 2021).

Next, with respect to offtake purchase agreements, which are also sometimes referred to as offtake contracts, these are agreements in which a certain amount of the product that a particular private company's project or facility produces is sold to usually a local utility or other public agency or entity such as a local electric membership co-operative at a previously agreed-upon price (Yescombe, 2013). A real-world example of this type of agreement would be a privately constructed solar farm that contracts with a local energy utility such as Georgia Power to sell all the electricity it generates in exchange for receiving a specific price per kilowatt-hour in return.

Third, with respect to input supply agreements, these are contracts in which a local municipality or governmental agency agrees to supply a private party's project with a certain amount of a particular type of input at a predetermined price for a set period of time (Yescombe, 2013). While not common in the United States at this time, an example of this type of agreement would be if a local municipality agrees to provide a private party with a certain amount of solid waste on a regular basis over the course of a year to be burned in its incinerator in order for this incinerator to generate a consistent amount of energy.

Lastly, in regards to the final two different types of agreements listed above—specifically construction contracts and operation and/or maintenance agreements—the contractual provisions that are typically included in each of these agreements are usually incorporated into the body of another contract such as a concession agreement or offtake contracts. However, in certain situations, there is sometimes a need for a separate stand-alone construction contract between the various parties involved in a particular public–private partnership, such as when different private parties are each responsible for the design, engineering, and construction of a proposed facility or project and the local municipality or governmental agency involved desires there to be one single private entity which is ultimately responsible to them for delivering a completed, turnkey project in a timely manner, instead of the public entity having to deal with each of these different private parties themselves (Yescombe, 2013). Typically though, it's not actually the local municipality or governmental agency which requests this separate stand-alone construction contract, but rather the proposed project's lender that requires it, and if the project is being financed in whole or part by issuance of municipal bonds instead, potential purchasers of these bonds will demand it as well (Yescombe, 2013).[5] Similarly, separate stand-alone operation and/or maintenance agreements are also necessary under certain circumstances, such as when a different private party will be operating the project or facility upon completion from the private party which originally designed and constructed it, or if multiple parties will share the responsibility and expense of maintaining a certain project or facility for a defined period of time after it has been completed (Yescombe, 2013). For example, a newly constructed toll road might have a very specialized and experienced operator to help ensure that future revenue projections are met, while the other private party who was responsible for this project's design, engineering, and completion, as well as likely future maintenance, would prefer to utilize its expertise to design and build another toll road elsewhere.

While the various types of agreements described above are typically the most common types of contracts governing public–private partnerships today, real estate development projects necessitate a much different type of agreement due to the complexity, cost, and lengthy timelines associated with the development process, which, depending on the jurisdiction and the legal professionals involved, could result in an additional agreement and/or contract also being required, if not more, such as in the Camperdown example that was previously mentioned. However, there are many different types of contracts and other agreements in existence today which govern the relationship between a private developer and a public entity such as a local city or county that are jointly involved in the development of real estate together, with some of the more commonly known specific types of contracts and agreements regulating public–private partnerships in this context consisting of FILOT and PILOT Agreements, Development Agreements, Economic Incentive Commitment Agreements, and Memorandum Of Understandings (collectively, "Development Agreements"). The following section will discuss several key contractual provisions to consider incorporating into any Development Agreement going forward, the majority of which should be permitted in most U.S. jurisdictions.

Key Contractual Provisions to Consider Incorporating into a Written P3 Agreement

Some of the key contractual provisions and information to consider incorporating into future written agreements and contracts between private developers and public entities such as a local municipality, or any subdivision thereof, which will also be interchangeably referred to as "Development Agreements" throughout the remainder of this chapter, consist of the minimum amount of capital the private party is required to invest in connection with a proposed real estate development project and the number of new jobs that are projected to be created by this project (including but not limited to, the approximate average salary or wages for these to-be-created employment opportunities), and by when this minimum capital investment has to be made and the expected new jobs created; clear deadlines for completing each major component of the project, if applicable, as well as the entire project as a whole, with automatic extensions of these deadlines upon the occurrence of certain unforeseen acts of god such as international wars, natural disasters, or a global pandemic; a detailed list of the exact incentives and other assistance, if applicable, that the public party will provide to the private developer for undertaking a particular proposed development project, and when, as well as any particular design features, public amenities such as a number of free parking spaces, and/or other public policy considerations such as the inclusion of a certain percentage of affordable housing units in the multi-family apartment portion of a proposed development, all of which the project is required to have in order to receive all the potential benefits from the public body that were previously identified. Additionally, several other key contractual provisions and other important information to consider including in written agreements and contracts such as these are as follows: quarterly and/or annual reporting obligations relating to the real estate development project which includes the total aggregate investment made in connection with this development as of the reports' respective date, the number of temporary construction jobs and new full-time employment opportunities, if any, this project has created thus far; comprehensive insurance requirements that explicitly set forth the minimum policy limits that will be permitted and the specific types of policies each party must obtain, including if any parties must be listed as an additional insured under one or more of the insurance policies that another party is mandated to acquire, as well as indemnity provisions to ensure that neither party is left on the hook financially due to the occurrence of an event that the other party was in the best position to mitigate, or possibly eliminate the risk entirely; and detailed recapture penalties and an enumerated list of different contractual remedies available to the non-breaching party due to the other party's failure to perform its respective obligations and responsibilities under the primary written agreement or contract governing the public–private partnership involved in the development of real estate, or one of the ancillary agreements and/or contracts between the same public and private parties relating thereto, if applicable.

While it is important for the written contracts and other agreements governing a public–private partnership formed for the purposes of developing real property to include as much of the contractual provisions and information identified in the preceding sentences as possible—whether it be through one agreement or multiple—ultimately the most important provisions and information to incorporate into any written agreement to be used in this setting is what is the private developer going to build exactly and when, and what are the specific incentives the public entity will provide to the private party for undertaking this development project and when, as well as what happens when one or more of the parties involved fails to perform as agreed-upon and the available remedies a party is entitled to for this breach of contract. A good real-world example of a written agreement incorporating many of the contractual provisions and pertinent information previously mentioned is the Economic Incentive Commitment Agreement traditionally used by Cobb County, GA when providing certain incentives which it has the authority to make available to specific businesses in order to encourage relocation to Cobb County or greater expansion within its jurisdictional boundaries, a form of which is included in Appendix A at the end of this book.

This is just a brief overview of some of the key contractual provisions to consider incorporating into future written P3 agreements and contracts concerning a new proposed real estate development project which members of the public and private sectors have joined together into a public–private partnership to make feasible, or even possible at all. Additionally, depending on which jurisdiction within the United States a certain proposed real estate development project will be built, there are also likely some certain state law provisions which will also need to be added to these written agreements and contracts, as well as other state-specific considerations. An attorney who is licensed in the particular jurisdiction where the proposed development is to be constructed—and actively practices in that same state—would be the best source of information regarding the state-specific provisions which need to be included in a written Development Agreement relating to the proposed real estate project to be built there, as well as likely provide the best legal counsel on other related issues that are specific to the particular jurisdiction in question and should be addressed in this written contract between the parties involved.

Other Notable Legal Issues to Consider before Entering into a Written P3 Agreement

In addition to some of the key contractual provisions and important information to consider incorporating into a written agreement or contract governing a public–private partnership formed for the primary purpose of developing a new real estate project, there are several other important legal issues to carefully think through before entering a formal Development Agreement, or even moving beyond just preliminary negotiations. First, from the private developer's perspective, formally entering into a written P3 agreement or contract with a public entity will require this private party to disclose a certain amount of information to a local municipality or governmental agency which is highly likely to be subject to an open records act, freedom of information act, or other state-level legislation or law obligating regular disclosures of certain information and activities by a local municipality or governmental agency to the general public, potentially causing private and very sensitive information such as the finances of the private development firm and sometimes it principals or partners, as well as anything unique and/or proprietary about the proposed project such as its design features, amenities, and future tenants that are still confidential, to get out to a significant number of people in short order and be easily accessible to many others. Second, from the public sector's perspective, in some jurisdictions there may be statutory requirements that a published notice (or possibly notices) regarding a Development Agreement to be considered by the local municipality or governmental agency be published in the legal organ of general circulation in the county where the proposed project is to be constructed, or posted on the public body's website a certain number of days before the meeting when it would be under consideration for approval by the public body in question.

Contracts in Context: The Case Study of Cobb County, Georgia's Tallest Tower

One example of a recent significant real estate development project in the State of Georgia which was made possible through public–private partnerships is the new North American Headquarters of TK Elevator in Cobb County, Georgia, which is located approximately 10–15 miles from the City of Atlanta and is in close proximity to the Atlanta Braves' Truist Park and The Battery Atlanta—a massive mixed-use development that surrounds the MLB franchise's stadium.[6] TK Elevator's new North American headquarters, which was finally completed in early 2022, is spread across multiple buildings in close proximity to The Battery-Atlanta, and includes over 200,000 square feet of leased office space and a separate state-of-the-art quality control tower for testing elevators that is approximately 420 feet tall—the tallest structure in all of Cobb County (Bruce, 2021). In total, the combined investment in TK Elevator's new North American headquarters between the Braves Development Company, LLC's special purpose entity responsible for developing the new office tower for the elevator company to occupy and TK Elevator's new test tower, tenant improvements, and other personal property to be located on site is estimated to be around at least $250,000,000 (Cunningham, 2018). To help lure this substantial amount of investment to Cobb County instead of it going to another municipality in Georgia or possibly another state all together, the Cobb County Board of Commissioners—the public body with the appropriate authority to act on behalf of the county—entered into a written contract known as an Economic Incentive Commitment Agreement by and between Cobb County, Georgia, Thyssenkrupp Elevator Corporation, Thyssenkrupp Real Estate North America, LLC (with Thyssenkrupp Elevator Corporation and Thyssenkrupp Real Estate North America, LLC collectively referred to as "Thyssenkrupp"), and BRED Co., LLC—an affiliate of the Braves Development Company, LLC, pursuant to which Cobb County authorized up to about $1.3 million in economic incentives for these private parties primarily through significantly capping permit fees—for example the building permit for the elevator testing tower should have cost Thyssenkrupp approximately $601,000, but was instead capped at only $5,000 (Cunningham, 2018).[7]

In addition to the financial incentives to be provided by Cobb County, GA through this Economic Incentive Commitment Agreement referenced above, the Cobb County Development Authority—a quasi-independent public body created for the purposes of, among other things, developing and promoting trade and commerce within Cobb County—approved a property tax abatement worth an estimated total of $15 million dollars through what is known as a "bonds-for-title" transaction due to the Georgia Constitution's strict prohibition on public entities within Georgia conferring a gift or gratuity on a private party, as well as its express uniformity of taxation requirement (Bruce, 2021).[8] In the simplest terms, a "bonds-for-title" transaction such as this involves a private party, or in this particular case three, transferring their property, buildings, improvements, and certain personal property and other equipment to the Cobb County Development Authority, which is not subject to taxation since it is considered a governmental entity, and then leasing it back from the Cobb County Development Authority during the period in which the property tax abatement is available, after which time the project will either automatically revert back to the respective parties who transferred property in fee simple or these parties can exercise a purchase option to acquire their property back for a small nominal amount of money.[9]

Ultimately, the presence of multiple written contracts and other agreements was necessary in in the case of TK Elevator's new North American headquarters due to the two different Cobb County entities which were offering incentives to encourage the development of this project, the various private parties involved in owning and developing different components of this overall project, as well as to comply with the Georgia Constitution which required the use of the "bonds-for-title" structure that is prevalent in the State of Georgia so that there was in fact a transaction of some kind, thus making the tax abatement the local development authority was going to provide permissible. Additionally, this cumbersome and complex process also likely provided the unexcepted ancillary benefits of helping the local community better understand the proposed development and creating a more exhaustive vetting process since proposed project had to go through two separate public approval processes.

Conclusion

In conclusion, public–private partnerships formed primarily for the purpose of developing real estate are becoming increasingly common, with more and more developers seeking some sort of monetary support, tax relief, and/or other assistance from the public sector with the projects they are currently working on since these benefits can be the difference in a project that is financially feasible, and one that is not.

While there are many different contractual provisions to consider incorporating into a Development Agreement, or the equivalent written agreement that is similar to this and utilized in your jurisdiction, the most important thing to remember with any agreement governing a particular public–private partnership engaged in real estate development is to make sure it is in writing, that the written contract or other agreement is detailed, but not overly complex or lengthy; and that there are no obvious ambiguities that could be interpreted differently by the respective parties involved or any other unclear or conflicting provisions contained in the written agreement between them, especially as it relates to material points and issues. A written P3 agreement or contract that is comprehensive in nature can also help safeguard both the public and private participants in a particular public–private partnership from future disputes and changes in local political leadership due to elections, while also providing clear deadlines as to when certain phases of the development project need to be completed and detailed recapture provisions that claw back any financial support the private developer may have received prior to he or she expending the minimum amount of capital investment that was previously agreed upon, or because the developer failed to create anywhere near the anticipated number of new jobs that were initially projected upon completion of the project.

Notes

1 Offtake purchase agreements are also sometimes alternatively referred to as "Offtake Contracts."

2 Specifically, according to a statement by GT Real Estate Holdings, LLC released to WCNC Charlotte on April 19, 2022, the special purpose entity behind the Carolina Panthers' proposed new development project claims that the City of Rock Hill, SC's deadline to fund public infrastructure was February 26, 2021, after which this entity notified the City of Rock Hill of its funding default on March 18, 2022 and the City failed to cure this alleged default within the 30-day grace period available to it (Eskieva & Shiff, 2022).

3 *Disclaimer*: All information and other material presented in this chapter are general in nature, and some points included herein may have been condensed and/or modified in the interest of readability, and do not, or are not intended to, constitute legal advice of any kind. Readers of this brief overview chapter are strongly encouraged to contact their own attorney or other legal counsel directly with any questions and/or to obtain legal advice regarding any issues they may be concerned about.

4 Specifically, as of October 23, 2022, the online version of the *Merriam-Webster Dictionary* says the term "public–private partnership" is not in the dictionary.

5 While potential purchasers of municipal bonds in the open market don't necessarily have the power to require a stand-alone construction contract that ensures one party has the ultimate responsibility for delivering a turnkey project or facility, potential bond purchasers will see this as a risk, and it will inevitably lead to higher interest rates and worse financing terms for the borrower.

6 TK Elevator was formerly known "Thyssenkrupp Elevator Company."

7 An unsigned draft of this Economic Incentive Commitment Agreement by and between Cobb County, Georgia, Thyssenkrupp Elevator Corporation, Thyssenkrupp Real Estate North America, LLC, and BRED Co., LLC is included in Appendix A at the end of this book.

8 Ga. Const. 1983, art. III, § VI, para.VI includes the prohibition on the Georgia General Assembly granting "any donation or gratuity or to forgive any debt or obligation owing to the public gratuity," while Ga. Const. 1983, art. VII, § I, para. III sets forth the requirement that subject to certain limited exceptions "all taxation shall be uniform upon the same class of subjects within the territorial limits of the authority levying the tax."

9 Due to the current language of the Georgia Constitution, the award of economic incentives such as a tax abatement from a local development authority must be structured as a "transaction" in order to comply with the State's Constitution so bonds are issued by the local development authority in most cases to a private party in the approximate amount of the project's cost in exchange for the local development authority taking title to the project during the abatement period and leasing it back to the private party for their use.

4 How Cities Use PPPs to Spur Real Estate Development

A Look at Greenville, South Carolina

Nancy P. Whitworth

Introduction

Greenville, South Carolina has a long history of public–private partnerships, which are most evident in the transformation of its downtown. Partners have worked in tandem to change the physical landscape and built environment to ensure Greenville's desirability for real estate investment. This chapter will focus on the approach Greenville, South Carolina took to stimulate real estate investments through the strength of effective public–private partnerships (PPPs). It is important to note that PPPs are not always just with the private sector; partnerships with philanthropic and not-for-profit entities were also incredibly important in many real estate projects.

Greenville's Background

Nestled in the foothills of the Blue Ridge Mountains along the I-85 corridor between Atlanta, Georgia and Charlotte, North Carolina, Greenville is one of the Southeast's fastest growing cities, with a current population of 70,000 within a county of over 540,000. Greenville has been blessed with a strong economy and in turn finds itself on a number of published "best of lists" across the country, a small sample of which includes the following:

One of the Best Downtowns in America, *Forbes Magazine*, 2011
America's Greatest Main Streets, *Travel & Leisure*, 2012
Best Small Cities in the US, *National Geographic*, 2018
8 Places with Waterfalls Right in the Middle of Downtown, *Livability,* 2018
Lessons in Human Scale Urban Design, *Toronto Globe & Mail*, 2019
Top Five Best Small City in the Nation, *Conde Nast Traveler*, 2021
One of the Coolest Small Cities in America, *Thrillest*, 2022

Greenville's nationally renowned downtown teems with vibrancy, livability, and commerce. Over 120 unique restaurants with sidewalk cafes and rooftop dining lie within a ten-block area with carefully curated public art interspersed. Within that same area are 15 distinct hotels and specialty retail. The performing arts center and

DOI: 10.1201/9781003222934-5

sports and entertainment arena provide world-class performers and entertainment. A variety of theater venues, museums of art, history, music, as well as a children's museum, round out some of the cultural offerings. Sports enthusiasts can enjoy a baseball game in a mixed-use stadium and entertainment complex, hockey in the sports and entertainment arena, and just a short distance from downtown take in a soccer game. Weekly concerts, major festivals, and events draw visitors and residents.

Well-loved public spaces abound, including Cleveland Park and Zoo; Falls Park on the Reedy, a 32-acre park in the heart of downtown and home to the pedestrian suspension Liberty Bridge overlooking the Reedy River falls; and Unity Park, a new 60-acre public park providing recreational experiences and a commitment to preservation of Greenville's natural resources. These major parks are linked via the Prisma Health Swamp Rabbit Trail, a 22-mile, and growing, multi-use greenway.

This current picturesque image of Greenville was not always part of its past.

Greenville was founded on a textile manufacturing economy, and up until the middle of the twentieth century prospered under that economy. Throughout the 1960s and 1970s, as the textile manufacturing industry declined, Greenville transitioned to a more diversified industry base with an international focus; however,

Figure 4.1 Reedy River Falls.

Greenville was left with a dying downtown in the midst of a growing region. Shopping centers lured major retailers to the suburbs and residents and businesses followed. By the late 1970s, Greenville's downtown was plagued with vacant storefronts, buildings in disrepair, and a major four-lane, parallel-parked Main Street devoid of any vegetation. Business leaders determined that a strong urban core was important to building Greenville's new economy and realized something dramatic had to occur. The plan focused on a transformative physical improvement which reduced the existing Main Street from four lanes to two lanes of vehicular traffic with free diagonal parking making way for canopy trees, lush plantings, and wide sidewalks. This set the stage to create a downtown atmosphere that would be conducive to office, specialty retail, entertainment, arts, and residential.

Greenville focused its plan on developing anchor projects through creative PPPs to spur redevelopment activity. This anchoring approach drove strategically located catalytic projects to fill community needs and maximize impact. Greenville's competitive strength in the 1970s was in its office potential and the realization that what makes a city a great place to live also makes a city a great place in which to invest.

The rebirth of Greenville's downtown is a case study in urban transformation. Its achievements have been the result of decades of commitment from public, private, and philanthropic leaders, propelled by an entrepreneurial spirit to take a very proactive role in making strategic investments and in working collaboratively to ensure quality development consistent with a community vision. These efforts have not gone unnoticed. Greenville has and continues to host city delegations representing over 23 states across the country who are interested in Greenville's story.

Greenville's Approach to PPP

The Urban Land Institute defines public–private partnerships as "'creative alliances' formed between a governmental entity and private developers to achieve a common purpose" (Corrigan, 2005: v). The operative words here are creative and common purpose. Each project is unique and requires creative approaches but all must align with a common purpose.

While structures and tools vary, there are some key ingredients of any successful public–private partnership, all which were at the core of Greenville's approach:

- Effective community leadership with mutual trust and respect between the private and public sectors;
- A collective vision that can be articulated into strategic actions;
- Innovation in financing and problem solving that combines the best of both the public and private partners;
- An entrepreneurial spirit with a willingness to take well calculated risks;
- A public advocate and go-to person within the public entity to champion the partnership; and
- Patience and commitment over a sustained period of time.

Effective Leadership with Collective Vision and Trust

Greenville's private sector has always factored heavily in its growth and prosperity. There was also a mindset that, for businesses to grow and prosper, community needs must also be met—it was the right thing to do, but it also made good business sense. Greenville's public sector understood its role that government alone could not ensure a prosperous destiny. Meshing the two sectors was best done when there was a shared vision articulated through a well-conceived plan.

Good planning is essential for PPPs to successfully deliver lasting results. Focusing on unique attributes of a place will help differentiate one community from another and create a product in the market that commands the highest value. Greenville transformed its downtown by creating a place that centers on the pedestrian experience and making it a place where people want to be. Attention to the natural environment, building design, signage, public art, and events all added to Greenville's identity. And realizing that cities must constantly reinvent themselves in order to prosper: plans were modified to adapt to changes through a transparent and participatory process. This provides predictability to the private sector and helps ensure that the community's expectations are met.

Mutual trust is not something that just happens. It takes work, desire, and time and must be grounded in an understanding of each sector's roles. The private sector is obviously looking for monetary returns on investment and the public sector's return is often in jobs created, taxes generated, or to stimulate other development, but both are looking for lasting value.

Innovative and Entrepreneurial Approach to Public Financing

The public sector must judiciously deploy resources to stimulate the desired private investment using incentives to appropriately meet the financial objectives of all.

In 1979 Greenville formed the Greenville Local Development Corporation (GLDC), a 501 c (4) not for profit to support economic development initiatives in the city. The GLDC provided the flexibility to provide loans, grants for infrastructure and façade improvements, invest as an equity partner, and use challenge grants to stimulate key economic development initiatives. Through an operating agreement, the city's economic development director served as chief operating officer. The board was independent from elected officials but remained in alignment with the city's economic development objectives to initiate opportunities that cities are often limited from doing.

Greenville used a variety of financial tools to support public–private partnerships, including the following.

Public Debt

GENERAL OBLIGATIONS BONDS

General Obligation (GO) bonds have the full faith and credit of the issuer and can be used for any public purpose, but debt capacity is limited. It was not the usual debt of choice for Greenville's PPPs.

REVENUE BONDS

Revenue bonds are limited to use for improvements of the revenue generating system such as parking, stormwater, sanitary sewer, hospitality, or tax increments.

INSTALLMENT PURCHASE REVENUE BONDS

Installment Purchase Revenue Bonds (IPRB) allow repayment from "any lawful source" and offer the most flexibility in terms of use and funding sources and are subject to annual appropriation so do not count against debt limits. The governmental unit must set up a corporation to hold title to financed assets and, as debt is repaid, fractional ownership of the asset is transferred to the city. IPRBs were regularly used in many of Greenville's development projects.

Financing Districts

TAX INCREMENT FINANCING (TIF)

Most states have varying legislation allowing the use of TIFs. In South Carolina, a TIF requires a defined redevelopment project area meeting conditions of blight or conservation. Investments in the redevelopment areas are required to be publicly owned and include a comprehensive plan and budget.

Beginning in 1986 and continuing through 2016, Tax Increment Financing was the most significant financing tool for Greenville's downtown revitalization. TIF bonds and revenues were used for public infrastructure— including streetscape improvements, public plazas, parking structures, property assembly, and public facilities. These critical investments leveraged a number of major transformative development partnerships.

SYNTHETIC TIFS

Synthetic TIFs became a public financing tool to address larger public–private partnership projects that did not fall within the original TIF boundaries. This tool offered more flexibility and was not subject to state enabling legislation. In a synthetic TIF, a developer pays for the cost of public improvements associated with a private development and is then repaid via an agreed-upon percentage of any new taxes generated. Cities do not incur any debt or up-front costs to ensure the new development. The only caveat is that the developer must be able to carry the up-front costs, and the city's due diligence should reveal all of the true costs of the improvements. Third-party verification of improvement costs and established caps up front provide an extra layer of accountability. Greenville initiated this approach in 2005 with Verdae Development, a 1,100-acre residential, commercial, and recreational planned development. The development entity constructed public streets and other associated infrastructure along with a major public park and were repaid over a period of 20 plus years with 25 percent of all new city property taxes generated. Subsequent projects in Greenville have used this same approach with negotiated terms.

MULTI-COUNTY BUSINESS PARKS

Generally used to encourage industrial investments, multi-county business parks can also be used in many states to provide property tax relief for projects that meet community objectives, including multi-family, mixed-use, and office development. In South Carolina, counties provide the designation of properties, and, if within a municipality, the municipality must consent. Once a property is designated within a park, the county can enter into a Fee in Lieu of Tax (FILOT) and a Special Source Agreement with the developer that allows the reduction of property taxes for a defined period. Ideally the reduction of taxes should be deemed necessary to solve for financial viability of a project and meet beneficial goals for the community.

Greenville used this incentive most recently in a downtown market rate multifamily mixed-use project by providing a 20-year, 50 percent FILOT in exchange for 20 percent affordable and workforce housing units. When using this incentive, it is important to ensure the affordable units are equally distributed over the type of units and the income levels of the tenants are verified initially, and periodically thereafter.

Accommodation and Hospitality Funds

Most states have an array of hospitality (taxes applied to prepared food and beverage) and accommodation taxes (taxes applied to hotel rooms). While these funds must have a nexus to hospitality, a variety of projects can fit this category.

Greenville tapped into these funds for projects such as hotels, a sports and entertainment arena, a baseball stadium/mixed-use development, parking garages, public plazas, and parks. After years of failed referendums to finance a coliseum, a PPP was formed to bring an arena to downtown. Accommodation taxes were an important component in making this work. The City and County approved a new 3 percent accommodations fee with 2.3 percent going to service debt for a new arena and 0.7 percent assigned to the Greenville Convention and Visitors Bureau for additional promotions.

Parking

Parking has been one of Greenville's key downtown incentives, especially in driving office development. The high land value in downtown coupled with the added cost of structured parking caused downtown to be less competitive as compared with suburban locations. From a planning and placemaking perspective, it was important for public parking structures to replace surface parking, have an enhanced level of design, and be placed in strategic locations to maximize usage.

Parking financing included various combinations of GO debt, TIF debt and revenues, parking revenue bonds, hospitality and accommodation taxes, and other grants when appropriate. A number of Greenville's public parking garages were constructed by a private developer as part of the PPP agreement to leverage new office, hospitality, and early residential projects. The city and development team worked closely in the design and costing to verify and satisfy city procurement costs. Payments were then released to developers on an agreed upon schedule.

Tax Credit Programs

NEW MARKET TAX CREDITS

The New Market Tax Credit (NMTC) program is an important tool for financing commercial real estate projects in designated low-income census tracts. NMTCs allow taxpayers to receive a credit for seven years against federal income taxes for making Qualified Equity Investments (QEIs) in private Community Development Entities (CDEs). The investments must be used in designated low-income communities and in projects that provide measurable community impacts and benefits.

New Market Tax Credits were crucial to a number of development projects in Greenville. A partnership was formed between the not-for profit Greenville Local Development Corporation (GLDC) and a private partner to form Greenville New Markets Opportunity (GNMO), which received an initial $80 million allocation in 2006. A number of key catalytic projects would not have happened without this valuable incentive. In recent years, service areas were expanded to the state and region, additional allocations were received, and a rebranding to The Innovate Fund was made. This positioned Greenville for continued NMTC investments and a revenue stream for other strategic investments to support city economic development.

MISCELLANEOUS MUNICIPAL, STATE, AND FEDERAL INCENTIVES

Local and state incentives vary but many provide tax credits for redevelopment of textile buildings or sites, abandoned buildings, historic properties, retail revitalization. A South Carolina state incentive for historic property renovation can freeze the property fair market value to pre-renovation for a period not to exceed 20 years, providing a valuable incentive for real estate projects. A state incentive for abandoned buildings has also been especially beneficial also providing tax credits for renovations. Federal historic tax credits are yet another incentive.

Early in the redevelopment of downtown, Greenville partnered with a consortium of banks to provide low-interest loans for redevelopment. In later years, Greenville utilized façade grants to encourage building upgrades in designated corridors. City staff became an important facilitator to ensure that developers and businesses were aware of and able to utilize any local, state, or federal incentives.

Other Value Adds

Value to the private development process is not always just about financial incentives. There are other ways the public partner can bring value to the partnership and to the project's bottom line, such as expedited reviews and permit approvals, flexibility in zoning requirements, single point of contact who can shepherd the private project and seek timely problem resolutions, building codes feasibility analyses to reduce surprises, facilitation of construction staging in use of public streets and rights of way, and ensuring that appropriate city staff are part of the development team from the initiation of a project. Time is money in any construction project and the more that the public can reduce time and provide predictability in its processes, the more likely the project can meet its desired targets. A true partner works to address problems during the development period and well beyond. Greenville never looked at a real estate project as a one and done, but rather as the beginning of a continuing partnership for the future.

A Public Advocate and Patience

A public advocate to serve as the go-to person and guide developer was an important factor in Greenville's success. The city's economic development director and team, which included community development, planning, and building codes, provided a coordinated and cohesive approach. Long-term stability in their roles provided consistency and a long-term view. Partnerships require patience, without which many of Greenville's signature projects would not have come to fruition.

Case Studies

Greenville has many examples of PPPs; it is quite simply how Greenville does business. Following are some brief case studies of catalytic real estate projects in downtown, as well as throughout the city, featuring examples of the different structures and incentives used throughout the years. It is important to note that real estate projects alone do not ensure success from a city's perspective. It is what happens within the buildings and surrounds that defines whether a place is and remains desirable to live, work, and play.

Greenville Commons

Need/Background

With physical improvements underway, Greenville looked to its strength in office to identify an anchor project which could initiate a downtown renaissance, and support and grow the office base. A new hotel and convention center were seen as the right fit leading to downtown's first major PPP and anchor. Underutilized and vacant properties were acquired directly on Main Street encompassing almost an entire city block and near potential office sites.

Project Description

Opening in 1982, the project included a 327-room Hyatt Regency hotel, 43,000 square foot (SF) of meeting and conference space, 100,000 SF office and retail, a 540-space parking garage later enlarged to 772 spaces, and public atrium and plaza.

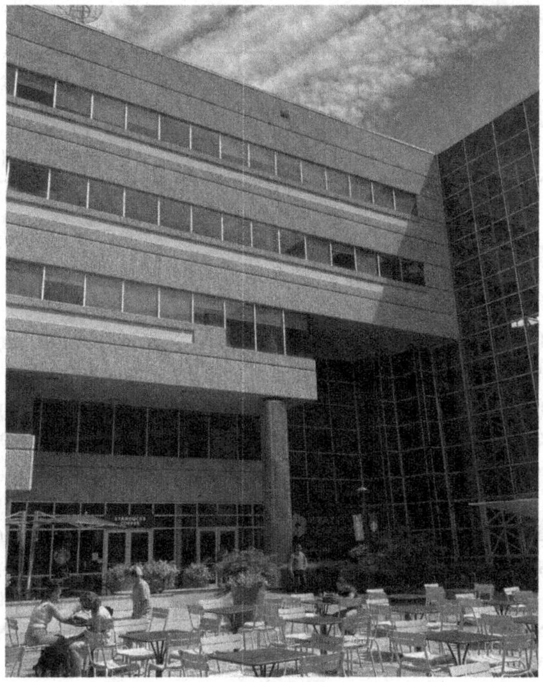

Figure 4.2 Hyatt Hotel and office.

PPP Structure

The city agreed to purchase the land for the complex and build the convention center, public spaces, and parking garage. The city leased the air rights to the hotel and office and also leased the convention center to the hotel. Some creativity was needed for the city's contribution, and the hotel atrium was considered a park. The private partners included the hotel and a group of local businessmen who invested at risk in the project, agreeing to a payment structure from cash flows, without which the project would not have happened.

Impact/Lessons Learned

While returns on cash flow may not have generated the most optimistic returns, the city benefited from the lease payments and parking revenues. More importantly the success of the hotel led to the construction of two adjacent office towers totaling over 445,000 SF and signaled the beginning of Greenville's renaissance.

Through another PPP, a local hotel group later acquired the hotel and city interests, retained the Hyatt Regency flag, and undertook a major renovation opening in 2013. The private plaza, though privately owned, is open to the public and hosts a variety of events and concerts.

Financing/Cost Layout

Total Project Cost	$34 million	Completed in 1999/2000
Private:	$18 million	Hotel
	$ 6 million	Local private investors
Public:	$10 million	HUD and City funds

Peace Center for the Performing Arts

Need/Background

In the late 1980s, a report commissioned by the local chamber identified Greenville's weakness in the area of culture and art. In 1989 Land Design/Research, Inc., through its *Downtown Development Plan and Program for Greenville,* identified a new vision for Greenville for the year 2000—"A thriving downtown which is recognized nationally as an example of a 'state-of-the-art' community in which to live, work, and play . . . which serves in itself as a national attraction" (1989: 8). With that challenge, the community stepped forward to begin work on a significant performing arts complex. Following Greenville's anchoring approach, a site was chosen for the Peace Center for the Performing Arts that could stabilize a deteriorating former industrial area and link downtown to a hidden asset—the river, waterfall, and park.

Project Description

Completed in 1990, the performing arts complex included a newly constructed 2,000-seat concert hall, a 400-seat theater, and an outdoor amphitheater. Five historic buildings on the site were preserved as part of the complex housing restaurant, office and retail uses.

PPP Structure

The PPP was initiated by a generous lead gift from the Peace family. Additional individual and corporate giving followed. The city acquired the necessary property for the project including business relocations, and also funded significant landscaping, amenities, and grounds maintenance. The city's funding was primarily done via tax increment financing (TIF). The county pledged accommodation taxes; and the state provided grants. The entire community provided support in a myriad of ways, demonstrating that success can lie in public/private/philanthropic partnerships when a common vision is shared.

Impact/Lessons Learned

The Peace Center became the cultural center of downtown and a key lynch pin in Greenville's downtown revitalization.

Figure 4.3 Peace Center.

The limited footprint of the site did not allow for any onsite parking, which turned out to be a good thing. Often cultural and entertainment venues make it too easy for patrons to come to a performance, enter and then leave, diminishing the economic impact for other business activity. Surrounding parking throughout the area was used, encouraging before and after performance patronage of local restaurants and retail thereby enhancing the economic impact of this asset. The PPP continued with a major reinvestment of over $22 million in 2010 as well as another announced for 2022, which continues to ensure the continued catalytic impact.

Financing/Cost Layout

Total Project Cost:	$42+ million	Completed in 1990
Private:	$28+ million	
Public:		
City:	$6.4 million TIF funds	
County:	$1.25 million Accommodation Taxes	
State:	$6 million grant	

Westin Poinsett Hotel/Poinsett Plaza Office

Need/Background

Infill and rehabilitation projects often require coordinating various developers in order to successfully achieve a desired outcome. Such was the case with a beloved threatened historic hotel in the heart of downtown. The hotel had ceased to operate and had been converted to senior housing. Lack of maintenance forced a closure and the once grand hotel began a precipitous decline. Efforts to convert to a residential use were not favorably viewed by the city. The renovation of the Poinsett Hotel back to a hotel was a lynch pin in Greenville's downtown revitalization plan. That plan also included a new office building and parking garage.

Project Description

The hotel's owners were not in a position to maintain their asset and further deterioration was threatening its survival. The city began to aggressively seek purchasers who could restore this landmark and return it to a prominent hotel. A number of ideas were floated, but none that would achieve the desired result. An out-of-town financier approached the city with a plan to finance the new city parking garage. The city indicated an interest in proceeding with the garage, *if* the Poinsett Hotel was purchased and renovated. The financier proceeded with a purchase of the hotel. Fortunately, a hotelier who had just completed an almost identical historic hotel renovation in Charleston agreed to acquire and renovate the hotel. A bank located in an historic building on the same block desired to expand and brought in a local office developer to construct a new office building. A church located in the same block sold its educational wing to yet another developer who developed 37 residential condominiums.

PPP Structure

The project required significant participation by the city in not only providing funding for the parking structure, public courtyard, and plaza for the hotel and other infrastructure, but also in balancing the multiple contractors and interests on a very tight site. Utility and servicing needs of the hotel and office buildings also required some creative solutions.

The PPP included:

Agreement with developers to construct a parking garage on behalf of the City, primarily funded with a TIF bond. Additional city funds were required for demolition of an existing private parking structure. The design of the parking garage included three unique sides to complement the hotel, office, and condo renovations.

Agreement with the office developer to assist in site demolition of a vacant department store, parking user agreements, and utility grants to facilitate complex utility requirements. The developer: constructed a new 220,000 SF office tower

Figure 4.4 Poinsett Hotel.

with four residential penthouses and street level restaurant space; renovated the existing historic bank building; and constructed a landscaped plaza with dining. Public art was donated to the city for placement in the plaza. The city used general funds and a grant from a local utility provider to address a complicated electrical servicing requirement and entered into user parking agreements for the tenants.

Agreement with the hotel developer for the complete renovation of the historic hotel and a new addition to better meet business traveler needs. Given the market environment at the time, additional support was needed to make this project feasible. The Greenville Local Development Corporation provided a fourth position loan with favorable terms and interest rate and closed the gap needed to take the project forward. Streetscape enhancements, public plaza and courtyard, and parking user agreements were part of the city's commitment.

Impact/Lessons Learned

The Poinsett project also began a new approach for the city in constructing public parking garages with the developer managing the design and construction as part of the development agreement. Using this approach, contractor conflicts and construction costs can be minimized. Communication and clear agreement on design, costs, and potential overruns are critical.

Financing/Cost Layout

Total Project Cost	$60 million	Completed in 1999–2000
Public:		
TIF Fund	$11.6 million	
City General Fund	3.5 million	
Total Public	$15.1 million	
Private:		
Hotel Investment	$19.0 million	Conventional financing including GLDC Subordinated loan Historic tax credits
Office Investment	$20 million	Conventional financing
Total Private	$39 million	

Riverplace

Need/Background

For years, property facing the riverfront of the Reedy River lay fallow. A local visionary leader, Tommy Wyche, began assembling a key ten-acre site located across from the Peace Center for the Performing Arts. Over a 25-year period, even though a number of developers expressed an interest in developing this site, Wyche had the foresight and patience to wait for the right time, developer, and project to realize the full potential of this transformational location.

Project Description

Wyche, representing the ownership partners, selected Hughes Development as the master developer. Ultimately five different developers would undertake various aspects of the project with a phased construction spanning over 15 years, with the last phase opening in 2016. The development includes an 87,700 SF office building, a 98,000 SF mixed use office with 27 residences; a 115-room Hampton Inn and Suites hotel; a 156-room Embassy Suites hotel, 54 residences in 3 separate buildings; street and riverwalk restaurants; a 626 space subterrain parking garage with artist studios occupying spaces along the riverwalk; public spaces and amenities including a man-made waterfall, interactive fountain, and carillon.

PPP Structure

Although the majority of the property had been privately assembled, additional property was needed to ensure that the public's access to the river was maintained in perpetuity. The acquisition was acrimonious and ultimately required eminent domain.

The city began working with Hughes Development and the development partners. The scope required a phased approach and coordination among several developers, each undertaking different construction projects. Given the multiple

Figure 4.5 Riverplace.

ownerships and the need to establish ground level connection, this partnership required layers of ownership over different elevations and a marriage of public and private interests—a first in Greenville.

The city entered into agreements with the developers to construct the parking garage and public improvements in three distinct phases. The land was donated in exchange for constructing the parking garage adequate to support the private vertical construction. Parking agreements were entered with the master developer and other users. The parking spaces fronting the river were exposed, so the city entered into agreement and master lease with one of the developers to convert these to affordable artist studios.

Impact/Lessons Learned

Riverplace was perhaps one of the most complex public–private partnerships for Greenville. It required patience, trust, and fortitude over a long period of time. There were many trying moments and inherent disagreements along the way, but somehow, through the relationship between the public and private partners, the project continued. The success of the physical project is evident by the myriad of uses all designed to fit appropriately and architecturally within one of Greenville's most sensitive environments. The public spaces are spectacular and became a standard for a true people place and spurred additional private development within the West End area of downtown. The city's returns transcended just the new property taxes generated by the buildings. Rather than a standalone parking garage, the city was also able to receive property taxes for buildings in the air rights over the garage. Accommodation and hospitality taxes along with business license fees were considerable.

Financing/Cost Layout

Total Project Cost:	$109 million	Completed in 2005–2016
Private	$ 90.6 million	Conventional financing
Public	$30 million	TIF bonds and funds, parking enterprise, hospitality tax, grants, local accommodation taxes, Sunday alcohol permits

One City Plaza (ONE)

Need/Background

As redevelopment continued, a glaring eyesore in the heart of downtown remained—a vacant Woolworths covering almost a city block. The store closed in 1991 shortly after signing a long-term lease. The property was also saddled with multiple owners spread out over the country and efforts to redevelop were stymied as there was no impetus for the owners to sell. An adjacent public plaza had undergone several renovations which never quite met the mark and hindered the street presence of an existing older office tower.

Project Description

The development included two offices towers (179,000 SF and 200,000 SF); 50,000 SF retail space; a 144-room Aloft hotel; a 475-space parking garage, redesigned public plaza and alleys, public restrooms, and a facelift to an existing 190,00 SF office tower. The development brought new national retailers such as Anthropologie and Orvis, along with other unique shops and restaurants.

PPP Structure

As the lease term on the vacant Woolworths began to expire, the city started working with a developer to assist in the acquisition, even obtaining a small ownership interest in the event a partition action was needed. After going through several developers, a local developer stepped into the project and began a multi-stage development program initially focusing on new office and street level retail.

Partnerships were structured between the city and the developer and also between the developer and Clemson University. Clemson had previously moved

Figure 4.6 ONE.

their Masters in Business Administration graduate program from the university's main campus 30 miles away to a leased downtown location. Growth in the program and a desire for Clemson to own their building led to this partnership. The developer provided the university with a gift-in-kind for 70,000 SF of shell office space. The city provided a favorable and flexible parking solution for the university and also entered into agreements with the office and hotel for user parking agreements. The city funded both the construction of a new parking garage and a complete redesign and rebrand of the public plaza, and transformed a connecting street into a pedestrian plaza and dining venue on one side of plaza and public restrooms on the other.

The developer donated the land for and managed the construction of the parking garage. The hotel was constructed above the parking garage, thus keeping this space on the tax rolls.

Impact/Lessons Learned

The ONE project reenergized the downtown core by bringing a mix of uses that drew both day and evening pedestrian traffic. New to market national retailers complemented existing retailers; new concept restaurants with outdoor dining enhanced the reconfigured and designed public spaces; Clemson's day and evening classes delivered a steady stream of students; and additional office workers and hotel visitors stimulated demand for retail and restaurants throughout downtown. Corporate office tenants were attracted to these new class-A buildings. This vibrant center was also able to attract an 80-year-old Greenville company to select the existing office tower as its corporate headquarters.

Financing/Cost Layout

Total Project Cost	$147 million	Completed in 2013–2015
Private:	$130 million	
	Financing:	Traditional Bank New Market Tax Credits State Retail Facilities Revitalization for former Woolworths SC Jobs Economic Development Authority
Public:	$17 million	
	Financing:	Installment Purchase Revenue Bonds with TIF, Parking, Hospitality funds

NEXT Innovation Center

Need/Background

Public–private partnerships take many forms that initially may not be just real estate focused. Such was *NEXT*, an initiative of the Greenville Chamber of Commerce in concert with the city and county of Greenville and other community partners with a mission to further cultivate the entrepreneurial climate and success for high-impact companies. In 2007, a group of entrepreneurs identified a need for economical and collaborative space close to the center of activity in downtown.

Project Description

Led by the entrepreneurs, *NEXT* provided the "software" and a partnership with a private developer that resulted in the *NEXT* Innovation Center, a 60,000-square foot renovation of a vacant distribution center. The space was custom tailored to meet the diverse needs of early-stage growth companies and became a convening space for *NEXT* members.

PPP Structure

A New Market Tax Credit allocation from The Innovate Fund (formerly Greenville New Market Opportunity) allowed the developer to keep rents affordable for these companies. The city's Greenville Local Development Corporation (GLDC)

Figure 4.7 NEXT.

also took an equity position in one of the major potential tenants. A major sewer upgrade by the city was also required. *NEXT* provided the daily hands-on support to the tenants.

Impact/Lessons Learned

PPPs are not just about providing financial incentives to the real estate project itself; but looking for other ways to infuse value such as assistance to a tenant. The GLDC's investment in a key tenant led to a good return on its investment and also garnered a new headquarters for the city. The success of this private model led to *NEXT* Manufacturing Center and *NEXT* on Main, similar privately developed and managed projects.

Financing/Cost Layout

Total Project Cost	$8.6 million	Opened 2008
Private	$8.6 million	New Market Tax Credits
Public	Support	Utility improvements
		Tenant support

Poe West

Need/Background

The Village of West Greenville, located less than two miles from the downtown core, is an emerging district of arts, creative businesses, and new residences, yet surrounded by neighborhoods with high unemployment. A major tract of land with an industrial use adjacent to, and comparable in size to, the entire Village of West Greenville commercial core became available. It was important to ensure that the redevelopment was sensitive to the surrounding neighborhoods and was complementary to commercial investments already made.

Project Description

Poe West encompasses over 65,000 SF in two separate buildings linked with a communal greenspace. Tenants include offices, fitness, brewery and winery, distillery, specialty foods and restaurant, and a culinary and hospitality innovation center.

PPP Structure

The idea for Poe West began in 2016 when the developers were looking for a real estate project that could provide jobs and educational opportunity, along

Figure 4.8 Poe West.

with a variety of collaborative businesses for the community. Searching for the right anchor tenant led to conversations with Greenville Technical College who had a robust culinary program but saw a new potential with the growing culinary and hospitality industry in the Greenville area. The result was the culinary and hospitality innovation center. The city's Greenville Local Development Corporation provided a challenge grant and a Truist bank provided the lead grant. The city entered into an agreement to provide reimbursements for enhanced streetscape and a variety of incentives for the project in order to keep rental rates low.

Impact/Lessons Learned

The Poe West project created a major anchor for this community by transforming an underutilized, but significant, piece of property into a vibrant mix of uses that complement and support businesses within the community. The project provides opportunities for training to support the hospitality industry and helps to address the unemployment in the surrounding community.

Financing/Cost Layout

Total Project Cost:	$12 million	Opened in 2020
Private:	$12 million	
	Financing:	New Market Tax Credits Historic Tax Credits, Special Tax Assessment, Brownfield Grant, Voluntary Cleanup and Brownfield Tax Credits
Public:	Financing	$175,000 for enhanced streetscape GLDC grant of $100,000 for the culinary institute

Conclusion

Greenville is a testament to when business, government, and the community work together, anything is possible; however, it is a process that requires constant cultivation and refinement.

Instilling a culture of partnerships within a public institution is not always intuitive. Greenville worked hard to understand, appreciate, and build the competency required for a successful real estate project. Greenville also worked with private partners to help them gain a better perspective on public processes.

Partnerships are often tested:

Pressure comes from all constituencies—elected officials, neighborhood groups, businesses, developers, and in PPPs all must be afforded an opportunity to engage.

Balancing transparency with confidentiality in a partnership can be a challenge.

Partnerships require mutual trust and respect, and relationships cannot be artificial. Regardless of how well agreements are crafted, something unexpected will happen.

Successful PPPs are not about "doing a deal," they are about coming together for mutual benefit.

Partnerships do not always work and can become litigious. Treat these as learning opportunities but never as future impediments.

5 Measuring the Costs and Benefits of a TIF-based Public–Private Partnership

Jeff Burton

Introduction

Each of the 50 U.S. states has individualized tax increment district (TID) statutes, court cases, and legal opinions for their jurisdictions. This model is based on Florida Statute 163, Part III, The Florida Community Redevelopment Act of 1969. The original law (HB-36A) was originally approved without a tax increment mechanism; this financial power was added in 1977 and then legally redefined by the State Supreme Court in 1980 (*The State of Florida v. Miami Beach Community Redevelopment Agency* (CRA)). The case modified the Act's Tax Increment Finance (TIF) into non-tax, increment revenue (IR) funds. This change allows a Florida CRA to bypass the State's Constitutional Article VII, Section 10's public funds for public uses requirement, leading to direct public–private partnership (P3) CRA to private sector engagement. The examples used in this chapter come from a hypothetical $100-million multifamily development in the City of Tampa, Florida.

To best understand this chapter's content and methodology, please read Chapter 2 in this book, "P3 Tools," which defines many of the variables used in this discussion. It also communicates my thoughts on the need for a new sustainable-development framework for future P3 investments and concerns regarding today's development model's negative anthropogenic outcomes. Also, no model is perfect, and as time passes, data, processes, and theoretical frameworks evolve. My work in this chapter is just a platform from which the reader can build. Where can you take it from here?

Why Is This Analysis Important to a P3?

This section discusses the importance of calculating a P3's short- and long-term costs and benefits. If a sustainable development framework (see Chapter 2) were applied, the project's economic, environmental, and social triple bottom line might also be examined for decisions leading to more holistic and healthy community outcomes. This framework, though not necessary to this chapter, estimates the opportunity cost synergies between the project's three perspectives.

DOI: 10.1201/9781003222934-6

Profitability

Economically, these estimates immediately create a "loss, breakeven, or profit" litmus test for the redevelopment authority that offers the TIF benefit. This analysis is useful as it predicts if the increment investment requested to incite private sector investment is economically feasible. This financial outcome also quantifies the costs of any societal and environmental added community benefits. In my anecdotal opinion, it is common for local governments to make unquantified and blind decisions when P3 investing. This causes a cashflow deficit as the public invests more than the project's TIF will return. Some agencies use a percentage of the annual increment to stay on the safe side of the bottom line. The TID's lifespan also plays a vital role in the equation.

The longer the TID lifespan, the more profitable the project may become to its authority, but at a longer-term "future value" loss to the private investor. This will be discussed later in the chapter. The converse is also true. Without an educated analysis, the public redevelopment partner may not invest enough and lose an opportunity for added private investment. These calculations should focus the TIF investment to maximize the community's goals. Does this equation determine if the development's local market is "primed" and needs the P3?

This mechanism does not determine the market's financial strength or actual developer's capital stack needs. Quantifying the value added to the development's taxable value for the short term, as well as the lifetime of the redevelopment authority, supports successful financial and political support. I created a cap-rate litmus test, measuring the local real estate need for this TIF-induced P3, but that is not covered in this chapter. Encouraging the private sector to create intrinsically (not profit-driven) valued community benefits, that cost financial profit, should be the driver behind these equations. These benefits should indirectly add to the real estate value over time, in the form of local resident job creation, reduced crime, healthy environment, historic preservation, and more.

Profitability is usually the developer's goal, and that's normal and acceptable. Without their profit-based decisions and investments, there would be no increment to analyze and invest in the first place! This is a self-funding partnership, generated by their motivations. Going one step deeper, the increment investment might be that one "puzzle piece" that encourages the risk-averse lender to invest in the developer's project. Finally, most private developers invest with a short-term profitability mindset, while public redevelopers plan with a long-term sustainable development strategy. Is this a problem?

Many redevelopment authority boards do not understand their agency's life mission and do not commit to the long game that only their authority can play. The redevelopment authority's increment is like a long-term investment that should not be decided by its board on a short-term return on investment (ROI). Many increment laws relate directly to slum and blighted area conditions. These conditions usually define communities with deflating tax bases and inflating public service costs. These conditions did not happen overnight and were usually encouraged by the local governing body's short-term financial strategies. The authority boards should be reminded to not expect a community-based long-term sustainable redevelopment result when they continue to try the same short-term ROI strategy. That is called insanity!

Slum and Blight

Blighted areas are known for their negative economic, social, and environmental characteristics. Economically, property values, commercial and residential rents, leases, and income bases may decline. Socially, crime, disease, building, code enforcement and zoning violations, especially density per unit, and special services like police, fire, emergency, and building and code enforcement may increase. Environmentally, brownfield sites may increase or have the perception of increasing, while water and air quality may degrade, and certain sensitive animal and plant life may fail and even cease to exist, while other vermin and invasive species may increase. This can lead to disease. In summary, blight is the downward trip a community makes on its way to slums.

Slum areas, though worse for the community, can become economically profitable, but for all the wrong reasons, and usually at a cost to the environment and society. Slums are known for overcrowded living units, rampant drug use, and other condoned criminal activities. They also promote high death rates, illegal dumping, and inhuman living and working conditions. In areas like this, the landlord becomes the slum lord. Florida Statute 163.340 (7) and (8) defines both “slum areas” and “blighted areas.” The State Legislature assigns 3 source elements to slums and 14 source elements to blighted areas.

Politics

Sometimes public decision-makers argue that redevelopment authorities take away from the whole community by reinvesting the area’s increment back into its district. This is usually a short-sighted political view for several reasons. First, if the redevelopment authority did not exist, the developer might not be incentivized to invest in the geographically slum and blighted areas, so there would be no increment to begin with. Second, if the redevelopment authority calculates and disperses the incentive along the estimates provided using this chapter’s methodology, the P3 is completely self-funding, and most likely generating a “public profit.” Third, the third-party public investor (city, county, or state) may also experience reduced service cost and greater financial profitability in the process, depending on how the law is constructed. Fourth, when the TID sun sets, its local governing authority receives back its entire value. For example, the Florida redevelopment law allows the redevelopment area’s investing taxing authorities to keep the base year value, and a 5 percent increment retainage (calculated and summed later in this chapter). Fifth, these dollars can be a significant revenue to the TID’s governing body, a sustainable revenue stream from an area that usually does not produce them. Generating these estimates is the best remedy to address politically motivated short-sightedness.

Monetizing

These quantifications also provide data-driven choices to the redevelopment authority for short- and long-term spending. Monetizing the projected increment revenue to fund larger community projects requires these quantifications to normalize the borrowing risk. What is the incentive provided to the developer, how is it determined, and how long is the term of payment? What is the present value of that incentive? These are questions that a third-party lender will ask when making a TIF monetization decision. As an added TID benefit, if the TID authority borrows against the estimated TIF to redevelop a public use such as roads and parks, third-party lenders can offer tax-free loan sources.

Intrinsic Value

Some P3 cases focus less on the economics and more on the intrinsic benefit to the environment or society. Brownfield redevelopments can push all these buttons. For example, the P3 remediates an environmentally polluted property, or one that is at least stigmatized as such. Second, many of these sites are in lower income minority neighborhoods. Addressing the contamination benefits the neighbor's health, environment, and economy. Third, besides the TIF benefits, the developer may also receive state tax credits for the remediation and possibly for low to moderate housing building materials (Florida offers these benefits). Historic preservation is another P3 case that can produce these intrinsic values. Also, in Florida, local governments sell properties at "real market value," but in the Florida Redevelopment Act of 1969 redevelopment authorities are allowed to sell properties, through a specific request for proposal, within their TID at "real value." This legal verbiage allows for historic preservation sites and other priceless properties, rather than just for profit. Let's take a look at a possible P3 project.

Project

Project Assumptions

The TID was created in 2006 and will be sunset in 2036. The city and county pay into its trust fund. The city pays 95 percent of the increment above the base year and the county pays 50 percent. The TID has a planning, building, and fire permit, impact fees, and a new resiliency fee rebate to be paid over its lifespan for new developments.

A 250-unit multifamily apartment development (no homestead property tax exemptions or property value annual increase ceiling) is planned within the TID. The development will be permitted in 2023, achieve its certificate of occupancy in 2024, and will be placed on the tax roll in 2026. The estimated permitted value for the project is $100,000,000, with a permit value of $500,000 and impact fees of $432,734.93.

The TID will commit $1,500,000 to be paid at 50 percent of the annual increment, starting in 2026.

What other data is needed? Key variables are as follows:

YEAR is the year value from the earliest millage value recorded to the TID's last one. This data set provides the timeline range for the forecast worksheet. Importantly, 15 data years of data is minimal to cast an accurate +/– 5 percent projection. City and county millage rates dating back to 1985 were located at the county Tax Collector's office. Apparently, this official was awarded the Florida Governor's Sterling Sustained Excellence Award for data efficiency. Year can also be a great longitudinal relational database key to other indirect data sets.

% GROWTH is the average annual growth rate of real property in the development's vicinity. These percentages may vary depending on the land use developed. This percentage may be discovered on the developer's project appraisal or requested as an estimate from the county property appraiser's office. Different land uses grow value at different rates. For this project, the state appraiser's association noted that the average growth in value for multifamily apartment projects averaged 3.1 percent growth over the past 20 years. This period included a hurricane and two economic recessions, which provides actual built in adjustments to the numbers.

Property Value

GROSS V is the value assigned to the development's estimated real property improvements minus the previous value of the site prior to the new development. This new dollar amount is estimated by the locality's building official, usually using the International Code Council's building valuation data, and found on the permit application or provided by the developer in the form of a licensed property appraiser's estimate. This estimate can be income or value based. Also, this number does not include tangible personal property. The previous value can be obtained from the county property appraiser or tax collector. At the beginning of the second year of occupancy, the equation looks like this:

GROSS V = GROSS VALUE + (GROSS V × % GROWTH)

DEP is the simple depreciation value of the real estate development. Every land use development has a lifespan. Other than the first year of its certificate of occupancy, each year the property ages, it depreciates in value, in contrast to inflation. The land is not included in the depreciation. For example, a multifamily project may have 27.5-year lifespan and be depreciated at 3.6 percent, starting in the second year.

NET V is the annual net value of the property once the development earns its occupancy.

NET V = GROSS V-DEP

Table 5.1 Millage and taxes

MILLAGE		TAXES		
CITY MIL	COUNTY MIL	CITY GROSS T	COUNTY GROSS T	GROSS T
6.540	9.304	$ -	$ -	$ -
6.540	9.144	$ -	$ -	$ -
6.540	9.056	$ -	$ -	$ -
6.540	8.931	$ -	$ -	$ -
6.540	8.816	$ -	$ -	$ -
6.540	8.691	$ -	$ -	$ -
6.540	8.734	$ -	$ -	$ -
6.540	8.504	$ -	$ -	$ -
6.540	8.454	$ -	$ -	$ -
6.540	8.192	$ -	$ -	$ -
5.733	7.729	$ -	$ -	$ -
5.733	6.853	$ -	$ -	$ -
5.733	6.886	$ -	$ -	$ -
5.733	6.882	$ -	$ -	$ -
5.733	6.886	$ -	$ -	$ -
5.733	6.882	$ -	$ -	$ -
5.733	6.876	$ -	$ -	$ -
5.733	6.836	$ -	$ -	$ -
5.733	6.784	$ -	$ -	$ -
5.733	6.755	$ -	$ -	$ -
5.733	6.728	$ -	$ -	$ -
6.208	6.693	$ -	$ -	$ -
6.208	6.661	$ -	$ -	$ -
6.208	6.635	$ -	$ -	$ -
6.208	5.731	$ -	$ -	$ -
7.475	5.731	$ -	$ -	$ -
6.180	5.594	$ -	$ -	$ -
6.153	5.457	$ -	$ -	$ -
6.126	5.321	$ -	$ -	$ -
6.099	5.184	$ 609,917.33	$ 518,403.26	$ 1,128,320.59
6.072	5.047	$ 603,492.99	$ 501,644.63	$ 1,105,137.63
6.045	4.911	$ 619,423.67	$ 503,186.32	$ 1,122,609.99
6.018	4.774	$ 635,762.09	$ 504,341.51	$ 1,140,103.60
5.991	4.637	$ 652,518.23	$ 505,084.75	$ 1,157,602.98
5.964	4.500	$ 669,702.28	$ 505,389.41	$ 1,175,091.69
5.937	4.364	$ 687,324.68	$ 505,227.57	$ 1,192,552.25
5.909	4.227	$ 705,396.07	$ 504,570.02	$ 1,209,966.09
5.882	4.090	$ 723,927.38	$ 503,386.17	$ 1,227,313.55
5.855	3.954	$ 742,929.74	$ 501,644.03	$ 1,244,573.77
5.828	3.817	$ 762,414.55	$ 499,310.13	$ 1,261,724.68
		$ 7,412,809.02	$ 5,552,187.80	$ 12,964,996.82

Table 5.2 Property value

		PROPERTY VALUE		
YEAR	% GROWTH	GROSS V	DEP	NET V
1997	0.00	$ -	$ -	$ -
1998	0.00	$ -	$ -	$ -
1999	0.00	$ -	$ -	$ -
2000	0.00	$ -	$ -	$ -
2001	0.00	$ -	$ -	$ -
2002	0.00	$ -	$ -	$ -
2003	0.00	$ -	$ -	$ -
2004	0.00	$ -	$ -	$ -
2005	0.00	$ -	$ -	$ -
2006	0.00	$ -	$ -	$ -
2007	0.00	$ -	$ -	$ -
2008	0.00	$ -	$ -	$ -
2009	0.00	$ -	$ -	$ -
2010	0.00	$ -	$ -	$ -
2011	0.00	$ -	$ -	$ -
2012	0.00	$ -	$ -	$ -
2013	0.00	$ -	$ -	$ -
2014	0.00	$ -	$ -	$ -
2015	0.00	$ -	$ -	$ -
2016	0.00	$ -	$ -	$ -
2017	0.00	$ -	$ -	$ -
2018	0.00	$ -	$ -	$ -
2019	0.00	$ -	$ -	$ -
2020	0.00	$ -	$ -	$ -
2021	0.00	$ -	$ -	$ -
2022	0.00	$ -	$ -	$ -
2023	0.000	$ -	$ -	$ -
2024	0.000	$ -	$ -	$ -
2025	0.000		$ -	$ -
2026	0.000	$ 100,000,000.00	$ -	$ 100,000,000.00
2027	0.031	$ 103,100,000.00	$ 3,711,600.00	$ 99,388,400.00
2028	0.031	$ 106,296,100.00	$ 3,826,659.60	$ 102,469,440.40
2029	0.031	$ 109,591,279.10	$ 3,945,286.05	$ 105,645,993.05
2030	0.031	$ 112,988,608.75	$ 4,067,589.92	$ 108,921,018.84
2031	0.031	$ 116,491,255.62	$ 4,193,685.20	$ 112,297,570.42
2032	0.031	$ 120,102,484.55	$ 4,323,689.44	$ 115,778,795.10
2033	0.031	$ 123,825,661.57	$ 4,457,723.82	$ 119,367,937.75
2034	0.031	$ 127,664,257.08	$ 4,595,913.25	$ 123,068,343.82
2035	0.031	$ 131,621,849.05	$ 4,738,386.57	$ 126,883,462.48
2036	0.031	$ 135,702,126.37	$ 4,885,276.55	$ 130,816,849.82

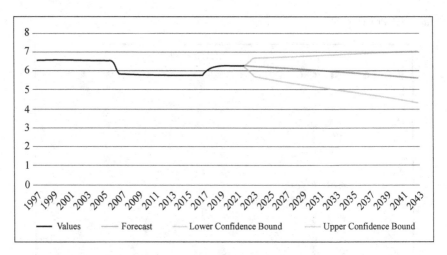

Figure 5.1 Millage.

MILLAGE

MILLAGE is the rate usually multiplied by the property's value divided by a denominator. For example, a property worth $1,000,000 in a city that has a 4.5 millage rate per every $1,000 of the property's value will pay $45,000 in property taxes that year. For this exercise, the millage rate for each taxing authority assigned to invest into the TIF trust fund must be discovered. For this project, *CITY MIL* and *COUNTY MIL* represent the annual millage rates for the taxing authorities investing in the TID trust fund.

To project this rate into the future and to the end of the TID lifespan (2036), past millage rates must be found. For best accurate results, a minimum of 15 data points should be used. In this case, the data was found back to 1985. Using the DATA tab on the Excel spreadsheet, FORECAST END is populated with 2036, TIMELINE RANGE is filled with DATE column from 1985 to the last factual millage rate (2022), and VALUE RANGE is measured with CITY MILL corresponding values. A forecasted projection is calculated for years 2027 to 2034. Populate the previous year's value should be subtracted from the new number to account for the existing increment prior corresponding years in the YEARS column. Repeat this forecast for COUNTY MIL. TAXES *CITY GROSS T* and COUNTY GROSS T represent the annual property taxes using the following formulas:

CITY GROSS T = (NETV/1000) × CITY MIL
COUNTY GROSS T = (NETV/1000) × COUNTY MIL

GROSS T is the sum of CITY GROSS T and COUNTY GROSS T. Please note that in the 11 years remaining on this TID, the estimated city and county property tax is $7,412,802 and $5,552,187 respectively. The total estimated taxes (+/− 5 percent) are $12,964,996.

Table 5.3 Millage.

Timeline	Values	Forecast	Lower Confidence Bound	Upper Confidence Bound
1997	6.54			
1998	6.54			
1999	6.54			
2000	6.54			
2001	6.54			
2002	6.54			
2003	6.54			
2004	6.54			
2005	6.54			
2006	6.54			
2007	5.733			
2008	5.733			
2009	5.733			
2010	5.733			
2011	5.733			
2012	5.733			
2013	5.733			
2014	5.733			
2015	5.733			
2016	5.733			
2017	5.733			
2018	6.2076			
2019	6.2076			
2020	6.2076			
2021	6.2076			
2022	6.2076	6.2076	6.21	6.21
2023		6.1804933	5.76	6.60
2024		6.1533867	5.64	6.66
2025		6.12628	5.54	6.71
2026		6.0991733	5.45	6.75
2027		6.0720667	5.36	6.78
2028		6.04496	5.28	6.81
2029		6.0178533	5.20	6.83
2030		5.9907467	5.13	6.85
2031		5.96364	5.05	6.87
2032		5.9365333	4.98	6.89
2033		5.9094267	4.91	6.91
2034		5.88232	4.85	6.92
2035		5.8552133	4.78	6.93
2036		5.8281067	4.71	6.94
2037		5.801	4.65	6.95
2038		5.7738933	4.59	6.96
2039		5.7467867	4.53	6.97
2040		5.71968	4.46	6.97
2041		5.6925733	4.40	6.98
2042		5.6654667	4.34	6.99
2043		5.63836	4.29	6.99

BASE

The BASE records the property value as it was on the day of the TID creation. For this project the land was vacant in 2006 and valued at $150,000. This number is harder to locate than most. Usually there is a sales price listed on the property appraiser's website or the tax bill from that year on the tax collector's site. In relation to the city and county property taxes, sometimes the property lines are different due to a subdivision or grouping. The important rule is consistency in reconstruction. You may use the average square foot price multiplied by the total area or add the subdivided parcels together. The CITY B is equal to $981, and the COUNTY B is $1,228. From YEARS 2026 to 2034 the columns will all be those two numbers. The TOTAL B is equal to $2,209 for rows. during the development's use through 2034, the city will collect $10,791 in BASE, while the county collects $13,508.

TAXING PERCENTAGE

The TAXING PERCENTAGE is equal to the percentage retained by each taxing authority from the TID trust fund. If you recall, the city committed 95 percent to the fund, while the county added 50 percent.

The CITY %, starting in 2006 is equal to 5 percent (100–95), while the COUNTY % is 50 percent (100–50). There should be a column for each taxing authority with their percentage per year.

TAX REMAINDER

The TAX REMAINDER is the value of the increment kept by the taxing authorities based on their TAXING PERCENTAGE. CITY $ starts in 2006 and is equal to 95 percent of CITY GROSS TAX, and the COUNTY $ follows suit.

CITY TAX REMAINDER = CITY GROSS TAX × CITY TAX PERCENT
COUNTY TAX REMAINDER = COUNTY GROSS TAX × COUNTY TAX PERCENT

From the estimate, it appears that the city will enjoy approximately $370, 640 from the development, during the TID lifespan. The county will retain $2,776,377.

Increment Revenue

This group of columns calculates the gross increment revenue from each taxing authority. CITY GROSS IR stores the result of the city gross tax minus the sum of the city base tax value and its taxing remainder.

CITY GROSS IR = CITY GROSS T − (CITY B + CITY $)

Table 5.4 Tax base

BASE			TAXING PERCENT	
CITY B	COUNTY B	TOTAL B	CITY %	COUNTY %
$ -	$ -	$ -		
$ -	$ -	$ -		
$ -	$ -	$ -		
$ -	$ -	$ -		
$ -	$ -	$ -		
$ -	$ -	$ -		
$ -	$ -	$ -		
$ -	$ -	$ -		
$ -	$ -	$ -		
$ -	$ -	$ -	5%	50%
$ -	$ -	$ -	5%	50%
$ -	$ -	$ -	5%	50%
$ -	$ -	$ -	5%	50%
$ -	$ -	$ -	5%	50%
$ -	$ -	$ -	5%	50%
$ -	$ -	$ -	5%	50%
$ -	$ -	$ -	5%	50%
$ -	$ -	$ -	5%	50%
$ -	$ -	$ -	5%	50%
$ -	$ -	$ -	5%	50%
$ -	$ -	$ -	5%	50%
$ -	$ -	$ -	5%	50%
$ -	$ -	$ -	5%	50%
$ -	$ -	$ -	5%	50%
$ -	$ -	$ -	5%	50%
$ -	$ -	$ -	5%	50%
$ -	$ -	$ -	5%	50%
$ -	$ -	$ -	5%	50%
$ -	$ -	$ -	5%	50%
$ 981.00	$ 1,228.00	$ 2,209.00	5%	50%
$ 981.00	$ 1,228.00	$ 2,209.00	5%	50%
$ 981.00	$ 1,228.00	$ 2,209.00	5%	50%
$ 981.00	$ 1,228.00	$ 2,209.00	5%	50%
$ 981.00	$ 1,228.00	$ 2,209.00	5%	50%
$ 981.00	$ 1,228.00	$ 2,209.00	5%	50%
$ 981.00	$ 1,228.00	$ 2,209.00	5%	50%
$ 981.00	$ 1,228.00	$ 2,209.00	5%	50%
$ 981.00	$ 1,228.00	$ 2,209.00	5%	50%
$ 981.00	$ 1,228.00	$ 2,209.00	5%	50%
$ 981.00	$ 1,228.00	$ 2,209.00	5%	50%
$ 10,791.00	$ 13,508.00	$ 24,299.00		

Table 5.5 Tax remainder

TAXING REMAINDER	
CITY $	COUNTY $
$ -	$ -
$ -	$ -
$ -	$ -
$ -	$ -
$ -	$ -
$ -	$ -
$ -	$ -
$ -	$ -
$ -	$ -
$ -	$ -
$ -	$ -
$ -	$ -
$ -	$ -
$ -	$ -
$ -	$ -
$ -	$ -
$ -	$ -
$ -	$ -
$ -	$ -
$ -	$ -
$ -	$ -
$ -	$ -
$ -	$ -
$ -	$ -
$ -	$ -
$ -	$ -
$ -	$ -
$ -	$ -
$ -	$ -
$ 30,495.87	$ 259,201.63
$ 30,174.65	$ 250,822.32
$ 30,971.18	$ 251,593.16
$ 31,788.10	$ 252,170.75
$ 32,625.91	$ 252,542.38
$ 33,485.11	$ 252,694.71
$ 34,366.23	$ 252,613.78
$ 35,269.80	$ 252,285.01
$ 36,196.37	$ 251,693.08
$ 37,146.49	$ 250,822.02
$ 38,120.73	$ 249,655.06
$ 370,640.45	$ 2,776,093.90

For the county, the same process is followed.

COUNTY GROSS IR = COUNTY GROSS T − (COUNTY B + COUNTY $)

The gross redevelopment authority increment revenue, or GROSS RA IR, is equal to the sum of all the taxing authorities' gross IR.

RA GROSS IR = CITY GROSS IR + COUNTY GROSS IR.

Over the 11-year lifespan, the city will contribute approximately $7,031,377.57, while the county amount is $2,762,963.47. Total IR for the redevelopment authority is approximately $9,793,963.47.

Net

ANNUAL % is the placeholder for percentage of annual increment revenue used to buy down the P# debt agreement. This case is 50 percent. TOTAL INC is the aggregate incentive amount agreed upon by the developer and the redevelopment authority. This case is $1,500,000. The DEVELOPER cells calculate the redevelopment authority's 50 percent payment contribution to the partnership and the CRA cells calculate the redevelopment authority's net annual increment revenue. This analysis set the contribution limits and predicts the number of years to payoff. This equation is built around an IF/THEN function to determine if the full amount of the incentive has been reached. IF the sum of all the prior DEVELOPER payments added to the RA GROSS IR multiplied by ANNUAL % is less than or equal to TOTAL INC, THEN RA GROSS IR is multiplied by the ANNUAL %. Else the sum of DEVELOPER PAYMENTS is subtracted from TOTAL INC. The Excel formula is the following:

IF
(SUM (PRIOR DEVELOPER) + ([@ [GROSS CRA IR]]
*[@ [ANNUAL %]])
IS LESS THAN OR EQUAL TO
[@ [TOTAL INC]]
THEN IF TRUE
@GROSS CRA INC*[@ [ANNUAL %]]
IF FALSE
[@ [TOTAL INC]] − SUM (GROSS CRA IR))

From the calculation, it should require four years to complete the developer's payment. The redevelopment authority will net approximately $8.2 million dollars that would not exist without the developer's investment to invest back into the community.

Table 5.6 Increment revenue

INCREMENT REVENUE		
CITY GROSS IR	COUNTY GROSS IR	GROSS CRA IR
$ -	$ -	$ -
$ -	$ -	$ -
$ -	$ -	$ -
$ -	$ -	$ -
$ -	$ -	$ -
$ -	$ -	$ -
$ -	$ -	$ -
$ -	$ -	$ -
$ -	$ -	$ -
$ -	$ -	$ -
$ -	$ -	$ -
$ -	$ -	$ -
$ -	$ -	$ -
$ -	$ -	$ -
$ -	$ -	$ -
$ -	$ -	$ -
$ -	$ -	$ -
$ -	$ -	$ -
$ -	$ -	$ -
$ -	$ -	$ -
$ -	$ -	$ -
$ -	$ -	$ -
$ -	$ -	$ -
$ -	$ -	$ -
$ -	$ -	$ -
$ -	$ -	$ -
$ -	$ -	$ -
$ -	$ -	$ -
$ -	$ -	$ -
$ 578,440.47	$ 257,973.63	$ 836,414.10
$ 572,337.34	$ 249,594.32	$ 821,931.66
$ 587,471.49	$ 250,365.16	$ 837,836.64
$ 602,992.99	$ 250,942.75	$ 853,935.74
$ 618,911.32	$ 251,314.38	$ 870,225.70
$ 635,236.17	$ 251,466.71	$ 886,702.87
$ 651,977.44	$ 251,385.78	$ 903,363.23
$ 669,145.27	$ 251,057.01	$ 920,202.28
$ 686,750.01	$ 250,465.08	$ 937,215.10
$ 704,802.25	$ 249,594.02	$ 954,396.27
$ 723,312.83	$ 248,427.06	$ 971,739.89
$ 7,031,377.57	$ 2,762,585.90	$ 9,793,963.47

Table 5.7 Net

NET			
ANNUAL %	TOTAL INC	DEVELOPER	CRA
		$ -	$ -
		$ -	$ -
		$ -	$ -
		$ -	$ -
		$ -	$ -
		$ -	$ -
		$ -	$ -
		$ -	$ -
		$ -	$ -
		$ -	$ -
		$ -	$ -
		$ -	$ -
		$ -	$ -
		$ -	$ -
		$ -	$ -
		$ -	$ -
		$ -	$ -
		$ -	$ -
		$ -	$ -
		$ -	$ -
		$ -	$ -
		$ -	$ -
		$ -	$ -
		$ -	$ -
		$ -	$ -
		$ -	$ -
		$ -	$ -
		$ -	$ -
		$ -	$ -
50%	$ 1,500,000.00	$ 418,207.05	$ 418,207.05
50%	$ 1,500,000.00	$ 410,965.83	$ 410,965.83
50%	$ 1,500,000.00	$ 418,918.32	$ 418,918.32
50%	$ 1,500,000.00	$ 251,908.80	$ 602,026.94
50%	$ 1,500,000.00	$ -	$ 870,225.70
50%	$ 1,500,000.00	$ -	$ 886,702.87
50%	$ 1,500,000.00	$ -	$ 903,363.23
50%	$ 1,500,000.00	$ -	$ 920,202.28
50%	$ 1,500,000.00	$ -	$ 937,215.10
50%	$ 1,500,000.00	$ -	$ 954,396.27
50%	$ 1,500,000.00	$ -	$ 971,739.89
		$ 1,500,000.00	$ 8,293,963.47

Policy

With all this chapter written, this section is the most important: codify your policies in writing and have them ratified by your authority. In some of the following examples, I use actual verbiage from a draft plan for the Community Redevelopment Agency. The policy may include the following sections.

Cover

Start with a cover that communicates what the policy is, the authority, date, and author, then add a page identifying the authority board, starting with the chair, members, advisory board members, legal team, and staff.

Table of Contents

The Table of Contents should identify each section designated for the policy. Let's take a quick glance at each. I like to write my policies like a building code, using numeric levels such as 1, 1.1, 1.1.1. This makes for easy referencing.

Definitions

The Policy should have definitions. Each term used should be italicized in the main content area, identifying to the reader that it has a specific definition. Do not skip this part, it is not important, until it is important. Use definitions from your state statute and other industry-approved documents. Clarification is good and helps with political guidance. The following is an example:

> Community policing innovation. A policing technique or strategy designed to reduce crime by reducing opportunities for, and increasing the perceived risks of engaging in, criminal activity through visible presence of police in the community, including, but not limited to, community mobilization, neighborhood block watch, citizen patrol, citizen contact patrol, foot patrol, neighborhood storefront police stations, field interrogation, or intensified motorized patrol.
>
> Florida Statute 163.340 Definitions (23)

Goal

The following identifies the policy's goal: "This program aims to remove and/or hinder Slum and/or Blighted Areas through private sector Increment Revenue grants in the Central Park Community Redevelopment Agency area and according to its plan." These types of comments are usually found in the TID plan and state statute.

Intent

Listing all the references to this document from the state statute and local plans conceptualizes the policy's intention as well as provides legal support. This alignment provides your redevelopment authority with political coverage when questioned why they support such a program. Also, this verbiage provides your board's attorney with the legal foundation needed to defend the policy if it is challenged. Basically, the intent section answers the question, "Why we did this?"

Eligibility

Who should and can apply? The incentives can be prioritized by the land uses, types desired, types okayed, and types declined. Eligibly can be tied to job creation, the used of minority, women, or veteran businesses, brownfield remediation, or any other measurable quantification found in your plans and policies.

Policies

This section discusses how the policy is managed normally and when something happens. Policies may include board makeup and management. When are meetings? How many attendees constitutes a quorum? Do virtual meetings constitute attendance? And more.

Requirements

What must the applicant provide to receive the P3 benefits? Is a permit card required? What about previous tax bills? What should the applicant expect to provide publicly to receive funds from the board?

Funding

This item can be displayed as a table that explains the financial provision for certain requirements. These items may be presented as dollar values or as a percentage.

Scope

This section defines the policy scope. This can include a geographic area, type of land use to be redeveloped, site remediation, and investment value.

Appendix

This chapter includes applications, maps, and refences.

Other Considerations

Net Present Value

Using a 4 percent interest rate, the net present value of the payments for this chapter's P3 equals $1.369 million. This value considers time in years times the cost (interest rate) of not having the money now.

It's an Incentive, Not an Entitlement!

What is the developer doing to earn the money? In other words, *do not just give them money*! Most redevelopment authorities have a working plan. What can the private developer do to help with that mission? The private sector can move much faster than the public side; they can also take actions that are not allowed to the public side. Create incentive policies with clear goals, vetted by your attorney, and ratified by your redevelopment board. Fort Myers, Florida has three measurable "community benefits" attached to their private developer incentives. First, the incentive is quantifiably tied to affordable housing based on state average median income prices for the area. Second, what percentages of the development subcontracts are performed by minority-, women-, and veteran-owned businesses? The authority recommends 15 percent. Third, developers are required to invest a value equal to 5 percent of the annual incentive in a TID-based non-profit like the YMCA, Boys and Girls Club, or Lee County Affordable Housing Coalition.

Self-Funding

If the developer does not invest, there is no increment. Let me repeat that. IF THE DEVELOPER DOES NOT INVEST, THERE IS NO INCREMENT. This should be at the top of your local authority's agenda, in bold, capitalized, and underlined. I have been asked by local leaders, "Why should we give them our money?" What money! It does not exist without them. This is why the Florida legislature wrote "to the greatest extent it determines to be feasible in carrying out the provisions of this part, shall afford maximum opportunity, consistent with the sound needs of the county or municipality as a whole, to the rehabilitation or redevelopment of the community redevelopment area by private enterprise" in F.S 163.345.

Profit Inflates the Tax Base!

TIDs are fueled by private development investment, and from the redevelopment authority perspective should be evaluated by their lifespans. As part of its contractual obligation to the developer, the redevelopment authority should demand that the developer not petition a tax value adjustment board to lower its tax responsibility. The contract should also require that the developer sell the incentivized development for a value higher than its current taxable value. Contractual fines should be included for developments that violate city codes and become a criminal activity source. Finally, language should be included that directs the authority when an entitlement changes hands. Does the incentive stay with the selling developer or transfer to the buying developer?

Death and Taxes

Do not be afraid of the taxes not being paid. Most states have created third-party tax-due investments. What does this mean? In Florida, the county tax collector lists all properties with late tax payments. These properties are digitally auctioned to the lowest return on investment tax certificate investors. If the same investor pays the taxes on a property three years running, the property can be foreclosed upon by that investor. Taxes will always be paid, creating increment revenue.

Conclusion

This Is a Proven Practice

This methodology is Florida specific and has been used to analyze over a billion dollars of potential private investment over the past ten years. It should be adaptable to other states and legislative regulations. Is it perfect? By no means! As a real estate development academic/professional, you are encouraged to make improvements and share them. Please send those amendments back to me, there is always room for improvement.

Respecting the Politics

Many redevelopment boards are seated by elected officials, and their authority must be respected. With that being said, a P3 professional's mission is to guide that board through the redevelopment process and that requires two things. One, knowledge of your redevelopment/increment revenue laws, what can and cannot be done, and what will and will not work. Ignorance is not a defense in the eyes of the law. Two, do not take it personally. Electeds think politically, if they say they do not, they are fooling themselves. It's the nature of the beast and there is nothing wrong with that (I was elected once and know!). Political thought is not the same as rational thought. You will be chastised and blamed, applauded, and rewarded sometimes in the same meeting! Take the good with the bad in the same stride. This takes us back to the first thing. The law will supersede politics, especially when the law is a governmental step higher than the political board. Knowing the redevelopment/TIF laws and communicating them tactfully is the best hand to play in this arena. To communicate to city, county, and state officials in a way that cuts through the political miasma. To be focused and intentional with questions and answers. To not "beat around the bush." To aim thoughts and words like a laser.

Data, Data, and More Data

Using data-driven analysis helps all stakeholders make informed dependable decisions. Local officials can estimate cost of incentives, terms, and net benefit to the community. Developers can prove the benefits of their investment in the community. The Community can look forward to TID improvements, benefiting local citizens. It's important to address all the stakeholders at the table, answering how each benefits from the development, and quantifying those amounts. Remember that, without the private investment, there is no public incentive created to offer.

Section 2

Public–Private Partnerships in Action

6 The Use of Public–Private Partnerships to Redevelop Greenville's West End through a Minor League Baseball Stadium Development

Stephen T. Buckman and James Frazier

Introduction

Developers by their very nature are entrepreneurial, and a key aspect of entrepreneurialism is taking advantage of situations that present themselves, especially those situations where a distressed party is involved and can be helped, or in more nefarious terms taken advantage of. This has been particularly evident in how developers have extracted incentives from cities that are in need of economic development. These incentives come in many forms from direct financial subsidies such as tax breaks or in the form of indirect subsidies such as streamlining the permitting process.

A subsidy that real estate developers are taking advantage of more prominently over the last couple of decades is the use of Public–Private Partnerships (P3). While this has been a key facet of how many cities have built infrastructure such as bridges, utility plants, and the like, for the real estate community this has been a relatively new phenomenon. But with cities in need of economic development, and with traditional incentives becoming hard to extract as cities face budgetary constraints and a political environment that does not see tax incentives in a positive light, P3 has become an important option. Thus, developers have looked to take advantage of urban economic issues by using the P3 process to their advantage.

The growth of P3 is a result of neoliberal policies of many urban regions. City governmental coffers, especially in the United States, are cash strapped at best and more often are on the precarious brink of one major catastrophe pushing them over the financial edge, as can be seen with the impact of COVID-19 on communities. Ironically this is true even as cities over the last few decades have grown in strength and prestige. With this reversal in fortunes of urban growth one would assume there would be a strong positive relationship with growth and financial strength. While the wealth of cities has grown and their bond ratings have increased, this growth has not been a one-to-one relationship; rather, urban growth has far exceeded urban coffers.

This uneven growth in population compared with the financial strength of cities is a direct result of the Reagan era of the 1980s, which made neo-liberal policies and the growth of neo-conservative urban administrations fashionable. These neo-conservative ideologies that were born of Reagan trickle-down economics swept

DOI: 10.1201/9781003222934-8

the country. The result of these policies was a reliance on the private sector to lead the way, and the government in turn was there to open paths for the private sector to succeed. The name of the game was low taxes, which created an environment to increase growth, especially in the sun belt cities of the South and Southwest, but at the cost of depleted financial coffers.

This urban economic situation is exemplified by the State of South Carolina and the city of Greenville. Due to an environment with low business and property taxes as well as anti-union laws, South Carolina was able to bring the North American headquarters of Bosch, BMW, and Michelin all in the Greenville area (recently Volvo open a massive plant outside of Charleston). With the close proximity of these plants, Greenville would be the beneficiary but to become so they needed to partner with developers to rebuild its infrastructure and create an urban area that was conducive to company executives. Considering they followed the path of State of South Carolina and many cities, the tax base coffers were thin at best.

The use of P3, which Greenville employed, has become the norm for communities rather than the anomaly. Cities more than ever cannot weather the economic storm and infrastructure and development of the built form without partnering with developers. While developers have often been the whipping boys of governments and community groups, the harsh fact is that many city governments simply would not be able to perform the public responsibilities to their constituents without developers. Thus, a d'état relationship often forms between governments and developers, each knowing that they need each other but both not necessarily liking it. Often going hand-in-hand with P3 approaches to growth is the notion of urban development regimes, which are often a grouping of power brokers that align to get a development built that more often than not benefits the regime first before the community.

This chapter will show how developers leverage P3, resulting not only in personal gain but also the redevelopment of a small U.S. southern city. To show this process we will use the case of the building of a Minor League Baseball stadium in Greenville, which helped to spur the growth of the city's West End and connected a linear downtown via the P3 development process that was put forth by the city.

Literature

There are three major theoretical areas of consideration that lay the foundation for the case study of Greenville. The three areas of inquiry include: public–private partnerships (P3): urban regime theory; and the idea of stadium development and the debates that surround it. Each of these points of inquiry shapes and explains the redevelopment of the West End and Fluor Field in Greenville.

Public–Private Partnerships (P3)

As touched upon, P3 is beginning to be considered more and more the norm of urban development, rather than an anomaly, and is manifested in real estate development and infrastructure among other avenues of public–private ventures. These partnerships are more the norm for various reasons, which include the fact the municipalities look to every development to benefit the community in some way. Developers are more wary of financial risks which municipalities can help control, and municipalities need more resources than they presently have to make projects they may do on their own more economically viable (Coomes, Burkland, & Fullerton, 2016).

In this regard, Delmon (2017: 2) defines P3 partnerships as development that "combines the strength of the public sector's mandate to deliver services and its role as a regulator and coordinator of public functions with the private sector's focus on profitability and commercial efficiency." In a more focused relationship to the real estate industry, the Urban Land Institute (the main global trade organization for the real estate development community) describes P3 as being composed of three broad areas: "a) to facilitate the development of a real estate asset to achieve greater benefits for both the public and private sectors; b) to develop and ensure the maintenance of critical infrastructure; and c) to design, build operate, and maintain public facilities, all in service of the goals of building sustainable, healthy, and resilient communities" (Coomes & Scheuer, 2016: 5). The thrust of P3 is the sharing of resources and the success of a development by both the private and public sectors. Thus, as Stainback (2000: 8) points out, in a traditional P3 "the public and the private partners structure a fair and reasonable sharing of costs, risks, responsibilities, and economic return."

For both the private and public sectors, P3 is highlighted by working together with the private sector providing their expertise of financing and efficiency with the public sector's ability to cut red tape and push developments through. Thus a P3 should provide: efficiency; whole asset life solutions: transparency and anti-corruption, technology, innovation, and know-how; and new sources of funding (Delmon, 2017).

There is an agreement that P3 should benefit both parties, hence the idea of a partnership. What then makes a successful P3? According to Corrigan et al. (2005), there are ten main principles to achieving success. These success factors range from preparation, a shared vision, rational decision making, and trust, to name a few. What can be seen is that P3 represents a chance for communities to use two distinct skill sets to get developments done that benefit the overall community.

Minor League Cities and Baseball

Minor league baseball (MiLB) offers an appealing avenue for smaller cities to jump on the bandwagon of stadium development in the name of generating economic growth. First and foremost, minor league teams and cities do not enjoy the de facto monopoly held by the majors; MiLB teams are often private subsidiaries of their parent major league team, which allows them to more freely move locations. The brands of major league teams are tightly linked to their home cities and drive a greater proportion of a team's value than minor league teams, which rely primarily on "fans in the seats" and cost-sharing with their parent clubs to profit. Additionally, minor league stadiums can be built for a fraction of the size and cost, typically holding generally no more than 10,000 compared with stadiums that seat 40,000 in the majors.

A minor league stadium's compact footprint allows for the stadium to be placed on smaller lots downtown and often results in little need for new parking structures, as the already established downtown parking infrastructure can typically handle the mostly night and weekend parking needs of roughly 70 home games a season (van Holm, 2019). The freedom of movement, business model, and smaller scale of minor league baseball allow for increased poaching across a larger pool of potential mid-sized suitors than the few dozen major metros that can reasonably support a top-level franchise.

With these parameters, MiLB stadiums are becoming important drivers. Yet even as their popularity continues to grow, their economic impact to communities is at best minimal (Agha & Ascher, 2016; Agha, 2013; Roy, 2008; van Holm 2018, 2019), with immediate jumps in attendance their first year and fading away by year five (Roy, 2008). The economic activity that does occur with MiLB is more geographically localized and is often a matter of shifting money and development that would have occurred in other areas of the community; this contrasts with major league baseball (MLB), which draws from a much larger geographic region with the greater potential to create new development (Agha & Ascher, 2016).

The more geographic-centered focus of MiLB was highlighted in a 2019 study by van Holm, which showed that there was an increase in home prices and new construction in census tracts near new stadiums relative to other areas in the city; however, when compared with sample tracts in cities that did not build a stadium, this difference disappears. This may play into the novelty aspect of new stadiums and suggests that MiLB stadiums are often beneficiaries, rather than drivers, of hyper localized development. Unlike their MLB counterparts, there appears to be little real economic development from MiLB stadiums. In turn, the true winners in these developments is not the community or the economic growth of the area but rather the urban regime of government officials, developers, and the team owners who spearhead the developments (van Holm, 2019). While the winners may be the regime, more often than not MiLB stadiums, while they have little economic impact on their own, can be economic redevelopment drivers, as in the case of Greenville.

A Brief History of Greenville

The city of Greenville is a microcosm of the history of twentieth-century America, from its highs as an industrial center to its lows as an example of the urban rust belt to postmodern metropolis based on tourism and the information economy (see Chapter 4 for more in depth overview of Greenville's growth). The city of Greenville, through the latter half of the nineteenth century and the first part of the twentieth century, was based on the manufacturing of the fiber to create clothing. In turn, Greenville and the greater upstate region became the epicenter of clothing manufacturing in the United States and was the epitome of a traditional milltown. Thus the mills ran and built the city into an industrial powerhouse. But, like much of the manufacturing base of the United States, they could not compete with cheap foreign labor, which resulted in the decline of the community.

For many years, the city of Greenville was a shell of itself, grasping at whatever it could do to spur economic growth. The fortunes of Greenville and the upstate changed with the decision in 1992 of BMW to locate their North American headquarters in the region. With the influx of BMW, Michelin, and Bosch, the city of Greenville, as the urban hub of the Upstate Region, now needed to create a downtown that was worthy of major industry.

The first measure the city took in the early 1990s was to clean and reinvigorate the Reedy River and Falls that ran through downtown and ironically was historically used as the engine of many of the mills. From the onset of the Reedy redevelopment, the city now needed people to move downtown, which was no easy task. Eventually, through aggressive P3 measures and incentives directed at developments, the first major development River Place (2008) along the Reedy River downtown took place, which became an overwhelming success. Not only was the development a success; so was the P3 that built the development. From this initial development the city has virtually exploded with development, with the newest one being the soon-to-come to market, mixed-use development of Camperdown (see Figure 6.1).

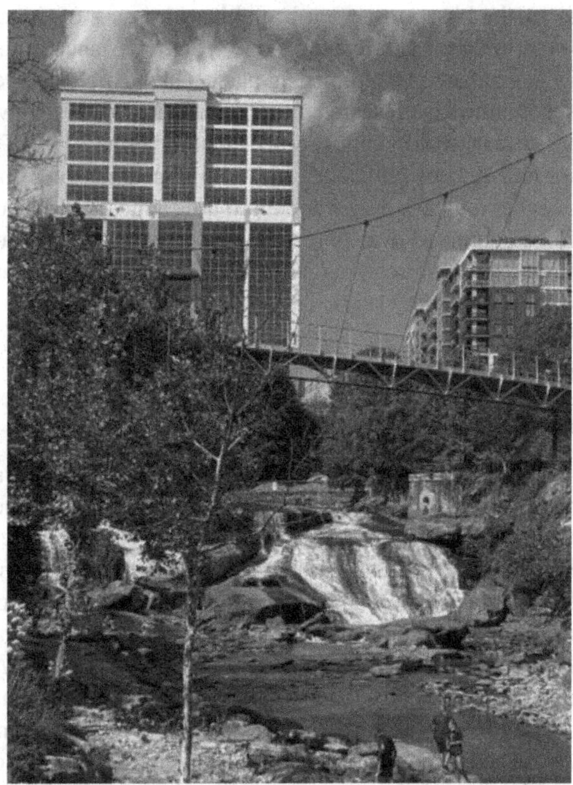

Figure 6.1 Falls Park with the new Camperdown Development in the background.

Thus, a city that 20 years ago was begging developers to come to their community is now fighting them off. Part of this maturation process was the completion of the linear path along Main Street that ran from the Hyatt on the north side, to Falls in the middle, to Fluor Field that was the southern terminus of the downtown Main Street corridor.

Methods

For this chapter we used a case study approach. This approach is traditionally used as a way to test theoretical models by applying them to real-world settings. This is accomplished using various normally more qualitative techniques such as interviews, ethnography, and in our case archival research (Cresswell, 2003). It should be noted that while the case study approach is traditionally a qualitative approach it can also be applied in a quantitative setting.

To construct our case study we utilized archival research, which included newspaper articles, journal articles, developer agreements, legal documents and city ordinances and resolutions, and web pages. Along with the archival aspect of the research, we engaged informal discussions with city officials and the developers.

As with any method of research there are pros and cons to the method used. The case study method has most often been hit with issues of rigor, replicability, and extrapolation, in that it only captures one specific instance in time and place, which is highly dependent on that time and place. On the flip side, the major pro of this method is that allows the researcher to view a theory in action, which can help to understand its "real world" implications (Shuttleworth, 2008). The case study approach is not meant to be replicable, rather it is meant to be a snapshot which allows the researcher to take a deeper look into how a theory plays out. Ideally one case study will be combined over time with similar case studies that examine the phenomena studied to construct a bigger picture, which shows the universality, or lack thereof, of how the theory studied acts in a real-world setting.

Case Study: Fluor Field

Predevelopment Process and Development Agreement

The development of Fluor Field took the form of a development agreement involving several different players. The City Manager at the time, James M. Bourey, entered into the development agreement on behalf of the city. While he took the position only a year before the baseball stadium discussions and development occurred, which successfully attracted the team to Greenville, he served long enough to see the project and partnership development prove successful in influencing the surrounding West End continued growth (Greenvillesc.gov, n.d., *History of City Managers*).

The development process began with the City of Greenville purchasing the land at the corner of South Main and South Markley Street. While the majority of the site was already owned and purchased from the Greenville County School System, an essential parcel to the development was under private ownership. Having identified the triangular-shaped parcel where Fluor field currently sits, the City entered into negotiations with the owners, Tad E. and Maureen M. Mallory. The purpose for the acquisition was defined as "to solidify the economic development program for the West End which allows for a mixed-use development, inclusive of a baseball stadium, on property fronting on Field Street and South Main Street" (*Resolution Property Purchase—Fluor Field*, November 22, 2004). The intentions for the project would prove to be much more than simply a baseball stadium. With plans not required to be finalized in order to purchase a property, the partnership sought to create a mixed-use development that would include retail and office space in an adjacent Field House building fronting Main Street. The mixed-use component would benefit the developer Centennial American Properties through securing the site. While they may have been able to acquire the site for the Field House project, the City's influence and ability to streamline the development process and timeline with the partnership already in place would benefit both parties, while also symbolizing the influence that the Greenville economic development department had in pushing the project through to fruition.

The parcel that needed to be purchased totaled approximately 0.9 acres. This portion was purchased from the individuals for $1,000,000 (*Resolution Property Purchase—Fluor Field*, November 22, 2004). While the purchase of this land could have easily been a "hold out" scenario, and the City could have had the power to take the land for "market value," as seen previously (see Chapter 4) through the eminent domain case for Greenville's Riverplace development site fronting the Reedy River, ultimately the power of eminent domain was used as leverage but not to implement. While there was no documentation that this was the case and the transaction was recorded as a standard property transaction, the fact remains unknown if the property could have sold for more in a healthy open market private transaction. The developer on the other hand potentially avoided community backlash by tying into the partnership and complementing a new baseball stadium with highly profitable Class A office space and high-end condominiums overlooking the outfield.

The full site sits at the corner of South Main Street and Markley Street, and is roughly 7 acres. The land acquisition costs totaled $4.1 million, which includes the land for the Fieldhouse mixed-use building to be developed by David Glenn and Centennial American Properties. In all, the development of the entire mixed-use site of the Fluor Field and the Fieldhouse office, retail, and condominiums would be a $29.2 million project.

The Partnership and Associated Funding

Fluor Field and the adjacent Fieldhouse project were built through two separate public–private partnerships, with the City of Greenville being the common member in both. With the ballpark, the Greenville "Bombers" proposal stated that it would pay for all associated stadium construction costs in order to secure both its spot as Greenville's minor league baseball team and the new stadium site. The Greenville Drive, under the entity RB3, LLC, would construct the 4,500-fixed seat stadium, berm seating for 2,000, a picnic pavilion, a playground for children, media board, and all other associated infrastructure and standing/walking areas inside the stadium, costing $15.5 million. Additionally, the City of Greenville would maintain ownership of the site and lease the property to the Greenville Drive (*City of Greenville—Fluor Field/Field House pdf*).

In addition to these costs, the specific responsibilities of each party are clearly defined through the development agreement and project outlay. Public costs in the project included the land acquisition and streetscaping, primarily along Main Street (*City of Greenville—Fluor Field/Field House pdf*). The streetscape was defined as adding an "attractive" combination of improvements, specifically including but not limited to lighting, landscaping bordering from the street to the walls of the development, and "other related and comparable improvements up to the boundary of the Developer Property" (*Ordinance_Fluor Field Development Agreement. pdf*). However, while the City of Greenville was required to provide two separate drainage and sewer access points, the streetscaping costs did not provide the capacity needed for the respective baseball stadium and Fieldhouse mixed-use project. These associated costs would be included in the private development and construction budgets. However, the Field House portion would need such infrastructure with or without the stadium, thus cutting the City's development costs on something the developer would have included in his budget regardless. Like any partnership, necessary communication between the partners is expected and important for successful planning and design. Therefore, the developer had input and control over the placement of these access points, despite not having to bear the burden of that access and extension cost (*Ordinance_Fluor Field Development Agreement. pdf*). Land acquisition and streetscaping totaled $4.1 million and $4.4 million respectively, bringing the public investment in the development to $8.5 million. The "Public Participation" in the Fluor Field was paid entirely by the City of Greenville (*City of Greenville—Fluor Field/Field House pdf*).

As a part of the Development Agreement, the City of Greenville would convey the property to the developer for the purpose of constructing the baseball stadium and adjacent mixed-use property, which was planned to front South Main Street and to activate the area of the West End through the entrance and retail storefront. The intention was that the baseball team and its stadium would support the restaurants, bars, and other retail and vice versa for both pre-game and post-game activity. Locating the stadium in an active downtown setting would draw fans to the area not only to catch a baseball game, but to enjoy everything that downtown has to offer before or after the game (*Ordinance_Fluor Field Development Agreement. pdf*). With the revamping of the area and the increased pedestrian traffic that the stadium would bring, parking and traffic control would need to be accounted for and planned for. This planning and associated costs fall on the public partner, the City of Greenville, as well. While parking would need to be provided, along with complementary landscaping, the City's expense for such parking would be capped at $100,000. There were two caveats, however, of the city providing parking. One was that the parking must be nearby and within walking distance of the stadium. While associated dedicating parking benefits the retailers, visitors, and the City, it also serves as a great amenity for the developer and property value. Located in an urban downtown, a submarket typically categorized by limited or undersupplied parking, the ability to provide parking to both office and retail tenants helps to attract customers—it likely can result in achieving higher rents or additional revenue from parking. The developer must keep the associated parking to service the development for ten years, which is to his benefit, or must otherwise reimburse the City of Greenville its initial investment for construction costs of $100,000 (*Ordinance_Fluor Field Development Agreement.pdf*). While the development of the site was expected, and is sure to increase the tax base of the site for the city, all partners and players recognized the significance of preserving and supporting the existing character and community that make up the West End Historic District, and more specifically the West Main Street corridor.

In the private investment for Fluor Field and the Fieldhouse, two players provided the capital for development and construction of the building and supporting infrastructure located on the site. These parties are the Greenville Drive, known as RB3, LLC in the associated documents and development agreement, and Centennial American Properties, under the leadership of local established developer David Glenn. While the two developments were separate, Centennial American Properties would act as the lead developer for the project, using the baseball club's equity to fund the cost of the stadium portion. This partnership between David Glenn and the Greenville Drive actually had long been in effect, with David Glenn and the owner of the team being close acquaintances. While some may construe this as twisted and an unfair deal situation, it perfectly exemplifies that real estate is a relationship-driven business. This relationship and associated trust was important in carrying the development to existence (*Ordinance_Fluor Field Development Agreement.pdf*).

The private investment requirements were outlined in the "Developer Responsibilities" section of the Development Agreement, driven by a minimum investment of $8 million from the developer. As clearly seen, the private investment would far surpass this minimum in order to create the neighborhood-transforming development. The minimum set an expectation for the quality of development and implementation of the investment to ensure that the City of Greenville's investment did not go to waste. The next requirement is compliance with zoning and building codes to make sure that the developer carries out the City of Greenville's vision and does not stray to a different use, particularly pertaining to the mixed-use component of the development. The third ensures that the developer implements that capital, setting a three-year time limit to spend $8 million on construction and improvements, or else will pay the City of Greenville liquidated damages for its investment in landscaping and infrastructure (*Ordinance_Fluor Field Development Agreement.pdf*). While the development agreement may appear to be a formality, as all three parties are invested in the project and the community, the recording of the documents is vital to ensure that each party is held accountable. The requirements insured that the City's vision for the site was not disregarded. While the developer had the majority of discretion in designing the project, the City primarily wanted to control the streetscape and first-floor retail that would activate the streetscape. In the end, the Greenville Drive provided an investment of $15.5 million for construction of the stadium, which was fully funded through future ticket sales and event income. Centennial American Properties provided an additional investment of

$5.2 million for the Fieldhouse. This mixed-use component of the development would include 40 residential condo units and 55,000 square feet of office and retail space (*City of Greenville—Fluor Field/Field House pdf*).

Last came the ground lease agreement giving the Greenville Drive the right to construct a stadium on the site. The parties agreed to a 20-year ground lease with an annual rent of $1. Essentially, the City of Greenville "gifted" the team the land to build a new stadium in the hope that the associated economic development would spur additional projects and revamp the surrounding neighborhood. The Greenville Drive was responsible for all construction costs, maintenance, operating costs and expenses, and any subsequent improvements to the site/stadium. Continuing with the vision of creating a vibrant and active corridor, the mixed-use portion's inclusion of retail was continually pounded home as the intention of the City when conveying the site to the developer (*Ordinance Lease in West End.pdf*). Additional investment and capital for improvement and expansion came in 2016. The agreement provided the Greenville Drive with a maximum contribution from the City of Greenville of up to $150,000 to fund improvements, as well as a total $5 million investment in the stadium. This funding would come from the City's eligible tourism-related funds and would be dedicated to the enhancements of Fluor Field and the stadium. This additional funding would not have been approved had the baseball team and the development not been a success through promoting and leading the growth of the West End Historic District of Downtown Greenville (*Resolution_Fluor Field 2016–36.pdf*, "Improvements to Fluor Field").

A Development Success Story

Fluor Field and the development partnership have become a large part of the overall success story of Downtown Greenville's urban revival and national spotlight. The stadium and team would bring new life to the forgotten textile neighborhood that lagged behind the rest of the city's growth. The public–private partnership would not only result in new excitement and activity to the West End Historic District but would also produce an award-winning baseball stadium. Its architectural design matching the character and feel of the existing community is not to go unappreciated, as old bricks were salvaged from abandoned and demolished textile mills to maintain the historic district's character, while also utilizing local resources. Additionally, public–private partnership would also be on full display. The City's funding would come from several resources, with a Tax Increment Financing (TIF) district being the most significant and the largest. Other financing came from West End Market sale proceeds, hospitality funds, and storm water and sewer funds (Whitworth & Neal, 2020). In examining the TIF district and the Fluor Field/Fieldhouse development's influence, the stadium is the reason that the TIF is significant. And as the team continues to pull in large crowds, other supporting retail, surrounding businesses and property owners, and streetscape infrastructure and landscaping would all be supported by the growing TIF funds (Smith, 2020).

The benefits to both the public side and private side of the development partnership are visible in the surrounding community. In first looking at the direct partners involved, the Greenville Drive has proven to be a success. In addition to its games drawing the largest baseball crowds in Greenville baseball history, the stadium is also entered into a regular rotation to host the Southern Conference Baseball Tournament, eventually hosting the tournament annually beginning in 2016 through at least 2019. The park also hosts several special events which include special occasion college baseball games such as South Carolina vs. Furman and several concerts that draw additional revenue to the stadium in addition to the Greenville Drive ("West End Events at Fluor Field", visitgreenvillesc.com). The stadium also hosts several events to benefit the local community, which include art events, the St. Patty's Day Dash and Bash 5k race, and several fundraisers and charity events (Worthy, 2020; see Figure 6.2).

Centennial American Properties owns a trophy mixed-use building that is central to the area and benefits from being adjacent to the baseball stadium through magnificent views. Consistently remaining fully occupied and stabilized, the commercial Fieldhouse development is a "home run" (Loopnet.com). The company maintains its primary office in the building, central to the neighborhood they were essential in transforming and enlivening into the active community it is today.

Following Fluor Field's success, the West End has created a unique identity in itself away from the primary downtown Greenville central business district (CBD), and continues to draw more visitors, residents, and interest from commercial and

Figure 6.2 Outside of the stadium and adjacent development.

residential developers. The project has been a driver of associated multifamily development, with The Greene completed across West Main Street from the stadium's entrance and .408 Jackson currently in development, which pays homage to Greenville local Shoeless Joe Jackson (Heather, 2019). The area has even begun to see hotel development take form recently, which is noteworthy as a new hotel market is created away from Greenville's CBD to benefit from business travel. Most recently, The Kimpton Hotel & Restaurants, a high-end boutique hotel, is underway and expected to open on Markley Street in 2022. While boutique hotel may not seem like a big deal, the Kimpton is a San Francisco-based chain with locations in high-end markets, which indicates that Greenville and its West End is "on the map" as a national destination (Connor, 2020).

Conclusion

The development Fluor Field is a steadfast example of how P3 works to get development done. The urban development coalition that was built solidified a development partnership that was able to work with the city to push through development. While the partnership faced little opposition to the development, which helped their development cause, the same cannot be said today.

The development that occurred around the stadium has resulted in rapid gentrification of the West End, which has led to the displacement of many disenfranchised communities and changed the community character of the area. The opposition that was more or less absent from the Fluor Field development has come front and center. Interestingly the activism around gentrification and development has taken two forms. First, the activism around displacement and whitewashing of a community which is being fought by clergy and a mostly displaced African American community on the one hand, and on the other hand wealthy home owners of the Alta Vista neighborhood who were the major force that fought the recent $1 billion proposed P3 County Square development in the West End.

The P3 that was established is one that became beneficial to both the city and the regime. In a nutshell the P3 worked the way it was supposed to work, boosting Greenville into the national spotlight as a case example for downtown rejuvenation. It will be interesting to see how successful development P3 schemes will evolve now that Greenville has become the "it" Southern downtown and is no longer desperate for development.

7 Innovation Districts and Misplaced Economic Development Incentives

Carla Maria Kayanan and Patrick Cooper-McCann

Introduction

In the aftermath of the Great Recession of 2008, many cities adopted a new strategy to promote redevelopment that purports to produce more equitable results: the innovation district. Innovation districts seek to spur economic innovation and job growth within designated zones by clustering high-tech firms in a mixed-use, "live–work–play" environment that features supportive infrastructure, like high-speed Internet and transit service, and supportive institutions, such as venture capital firms, businesses incubators, and research universities (Drucker, Kayanan, & Renski, 2019). Creating a successful innovation district requires a robust public–private partnership (P3): a structured collaboration among public, private, and nonprofit stakeholders to achieve a common objective (Sagalyn, 2007).

Urban leaders in the United States have long relied on public–private partnerships to accomplish two interrelated goals: to redevelop distressed neighborhoods and to promote economic growth and diversification (Beauregard, 1998). In the 1950s and 1960s, these goals were primarily achieved using federal funding for urban renewal, which enabled local governments to demolish and redevelop "blighted" properties in partnership with local business interests. Since the 1970s, redevelopment has depended more on place-based economic development tools that make it profitable for private developers to lead the redevelopment process. Growth coalitions have used such tools—including tax-increment financing (TIF), business improvement districts (BIDs), and enterprise zones (EZs)—to add amenities to urban neighborhoods and encourage job growth in emerging sectors, but the results have been uneven, socially and spatially. Innovation districts are developed using TIFs, BIDs, and EZs, but innovation districts are supposed to do more than physically upgrade a neighborhood and flag it as a promising location for investment. Because they focus on tech startups, innovation districts are expected to produce *more jobs* than other strategies. They are also supposed to produce new jobs in the service sector for lower-skilled workers, making growth *more inclusive*. Finally, because innovation districts emphasize live–work–play environments, they are also supposed to produce *more vibrant and sustainable cities* (Katz, Geolas, & Wagner, 2021; Katz & Wagner, 2014). Yet, innovation districts are still quite new, so the jury is out on whether these promises are being fulfilled.

DOI: 10.1201/9781003222934-9

This chapter discusses whether or not innovation districts are likely to produce equitable development. The chapter first explains how local officials came to rely on public–private partnerships and place-based economic development tools, including TIFs, BIDs, and EZs, to finance redevelopment—and why these tools have been subject to criticism. It then introduces the concept of the innovation district and examines the implementation of the concept in four locations in the United States. Through these examples we analyze how innovation districts compare with prior place-based economic development approaches and discuss the implications for real estate developers. The chapter concludes with a discussion of the advantages and shortcomings of this new concept compared with previous approaches.

The Pursuit of Post-industrialism through Public–Private Partnership

Urban leaders have relied on public–private partnerships to achieve their redevelopment goals since World War II. Industrial cities emerged from the war in deteriorated condition. During the war and afterward, manufacturing jobs decentralized as corporations reduced their inner-city workforces and built lower-cost factories on the metropolitan periphery. Concurrently, older neighborhoods were depopulating as white middle-class households took advantage of federal home mortgage subsidies and new interstate highways and relocated to the suburbs where they increasingly found work (Sugrue, 1996). Retailers suburbanized as well, contributing to the decline of neighborhood retail strips and central business districts (Loukaitou-Sideris, 2000). In response to this crisis, urban leaders adopted master plans to redesign inner-city neighborhoods to better compete with the suburbs for middle-class households and jobs, especially in growing sectors like government, education, real estate, finance, insurance, and research. These postwar plans kick-started decades of public–private efforts, which continue to the present, to revitalize central-city neighborhoods and industrial areas and maintain cities' economic competitiveness.

In the 1950s and 1960s, the federal government fueled redevelopment through the "urban renewal" program. This federal grant program covered two-thirds of the cost to acquire and demolish buildings deemed "blighted" or "obsolete" (Beauregard, 1998). After demolition, cleared sites could be transferred to real estate developers to be rebuilt according to a city-approved plan. Initially, urban renewal was used exclusively to redevelop housing, but after 1954, cities were allowed to pursue commercial and industrial redevelopment as well. Cities used the program to tear down blocks of older housing and replace them with university campuses, cultural centers, medical complexes, corporate and governmental offices, convention centers, sports arenas, and light industrial districts. These developments were typically built in a low-rise, modernist style, featuring ample parking and easy access to newly constructed interstate highways, just like the new corporate campuses being built in the suburbs. These developments helped advance a postindustrial transition, but they did so at great cost in a manner that harmed minority communities disproportionately, inviting a backlash against the "federal bulldozer." Tens of thousands of low-income and minority renters were displaced from their homes with little or no compensation (Ammon, 2016).

The federal government withdrew support for the urban renewal program in the early 1970s, despite accelerating trends of depopulation and job loss in older cities. The Reagan administration subsequently slashed local revenue sharing. Amidst a climate of fiscal austerity, in which the general operating funds for cities became insufficient, urban leaders were pressured to embrace the imperative of entrepreneurial management (Harvey, 1989; Mollenkopf, 1983). Unable to depend on significant federal or state transfers, local leaders were forced into competition to attract financial capital to their cities, pushing leaders to embrace a shared growth agenda with the private sector (Logan & Molotch, 2007). Public–private partnerships—agreements across a wide profile of urban leaders, including real estate developers, owners and managers of firms, politicians, labor unions, and the public sector—became the dominant development model (Sagalyn, 2007), with tax exemptions and deregulatory mechanisms undergirding the financing of urban and economic development schemes (Weber, 2002). Within this paradigm, urban leaders came to rely on a set of new economic development tools, including tax increment financing, business improvement districts, and enterprise zones, that enable the redevelopment of specific neighborhoods through partnerships with the private sector. State and federal governments still played a role in promoting urban redevelopment, but intergovernmental assistance came primarily through tax incentives rather than through grants.

Different economic development tools served different needs. Tax-increment financing enables a public authority to borrow money upfront to pay for infrastructure upgrades or other improvements to a targeted area with the expectation that these improvements will lead to higher property values, thereby enabling the authority to pay back the loan through increased property tax collection. If property values rise sufficiently, the loan "pays for itself" through taxes that would not have otherwise been collected if property values had not risen. TIF originated in California in the 1950s as a method for generating the local matching funds needed to receive federal urban renewal grants, but by the 1970s, many states had adopted it as their primary means of financing local redevelopment projects, including sports arenas, convention centers, and festival marketplaces, which were used to rebrand downtowns as destinations for tourism and entertainment (Briffault, 2010; Dye & Merriman, 2006; Weber & O'Neill-Kohl, 2013). TIF also became a popular financing tool for brownfield redevelopment projects, including factories and big-box stores (Coleman & Murphy, 2014).

Business improvement districts became a popular tool for revitalizing commercial districts. BIDs originated in Toronto in the 1970s and quickly spread to the United States (Houstoun, 2003; Levy, 2001; Mitchell, 2001). A BID enables the commercial property owners within a district to impose a supplementary tax on themselves. If approved by property owners representing a majority of property valuation in the district, all owners in the district must pay the tax. Rather than being deposited in the municipality's general fund, the revenue is returned to the business association to pay for programs and improvements to beautify and secure the district. Common expenditures include trash pick-up, park maintenance, signage and decorations, branding campaigns, street festivals, security patrols, graffiti removal, and "ambassador" programs to guide visitors. Unlike TIF, BIDs typically pay for everyday expenditures, rather than major infrastructure that requires taking out a loan. BIDs are an especially popular tool for revitalizing urban "Main Streets" in historic neighborhoods. They are also used to privately fund the management of parks and plazas (Seidman, 2004).

Enterprise zones were first developed in Britain in the early 1980s as a tool for encouraging business investment in underperforming and low-income areas. Enterprise zones target a mix of tax incentives, regulatory changes, and public investment to a designated area. The EZ concept was first implemented in Britain in the early 1980s and was quickly adopted in the United States, first at the state level and then at the federal level under President Bill Clinton (Boyle & Eisinger, 2001; Hall, 1982). The core idea is that growth can be stimulated in a distressed area by reducing the cost of doing business there; poverty can be alleviated at the same time if employers are incentivized to hire locally. A similar philosophy animates many nonprofit- and foundation-led revitalization initiatives, which seek to capture as much investment in one place as possible, including by capturing competitive federal grants and tax incentives, like the New Markets Tax Credit or the Opportunity Zones Tax Credit. Echoes of the concept persist in federal urban policy through targeted programs like the Obama administration's Promise Neighborhoods (Swanstrom, 2016). Enterprise zones may be used in conjunction with other economic development tools, including TIF and BIDs.

The aim of each of these approaches is to generate sustained growth within economically underperforming areas by making it more profitable for developers and business owners to invest. Cities have relied on these tools to facilitate a transition from an industrial economy to what is sometimes labeled the knowledge economy (Sassen, 2001; Scott, 2001). As the base of the economy has shifted, cities have had to find new uses for obsolete docklands, rail yards, warehouse districts, and factories, as well as the residential spaces that adjoin them. Pittsburgh is a commonly cited example of a city that has done so successfully through public–private redevelopment. Once dominated by the steel industry, Pittsburgh is today known for its green waterfront, corporate headquarters, research universities, hospital systems, and highly educated workforce, which makes the city a draw for both firms and "talent" looking to relocate from elsewhere. However, even success stories like Pittsburgh are highly unequal internally, with many working-class areas continuing to experience poverty and depopulation (Neumann, 2016). Furthermore, despite using similar tools, other industrial cities, like Flint and Youngstown, have experienced protracted decline. Unable to secure leading-edge employers like research institutions, technology firms, or banks, struggling communities settle for attracting less desirable developments, including casinos, waste processing facilities, and warehouses (Taft, 2018).

There are also risks and drawbacks to each specific economic development tool. TIF, for example, diverts revenue that might otherwise go to schools and other critical public investments in infrastructure, public safety, and civic governance, and it is challenging to prove that development in an area would not have happened if the TIF had not been proposed (Lester, 2014; Weber, 2002). This is a criticism that applies to tax breaks and subsidies more generally. TIFs have also been challenged on their ability to create employment opportunities. Because TIF is contingent on rising property values, TIF authorities tend to favor projects that will increase property values rather than create as many jobs as possible. Finally, TIFs have proved problematic to public coffers in instances when the increment is not enough to pay off the credit. If, for example, private investment fails to materialize or property values are stagnant, then the city may have to cover the losses to prevent the TIF authority from defaulting on bonds. In practice, over-enthusiastic projects are the norm, resulting in larger and mounting TIF debt (Pacewicz, 2013). For these reasons, California dissolved over 400 TIF authorities in 2012 and other states have considered following suit (Bieri & Kayanan, 2014).

BIDs and EZs have also fallen short of their promises. Business improvement districts pose less financial risk than TIF because they do not deprive local governments of tax revenue. Nevertheless, BIDs have been criticized for transferring control over the public realm to business owners. BIDs are also contingent on property values for success. Downtowns with high property values can generate substantial revenue through a supplementary tax, but the BID model is less useful in low-income neighborhoods where property values are low (Gross, 2005). Enterprise zones have also been criticized as being ineffective at alleviating poverty. EZs reward businesses with tax breaks for locating in low-income areas and hiring local residents, but relatively few EZs have made a significant difference with respect to decreasing inequality (Boyle & Eisinger, 2001). These drawbacks have left many cities seeking more equitable and effective approaches to redevelopment.

The Innovation District Concept

Since the 2008/9 Great Recession, some economic development theorists and practitioners have championed the innovation district as a new strategy for achieving postindustrial redevelopment utilizing existing tools such as TIF, BIDs, or EZs. Like prior efforts, the goal is to generate real estate investments and jobs in a targeted area. However, innovation districts are distinct because they emphasize high-tech and creative startups and high-density, transit-oriented, walkable urbanism (Drucker et al., 2019; Kayanan, 2022).

In the words of Bruce Katz and Jennifer Wagner—the two Brookings Institution scholars most responsible for popularizing the concept—innovation districts are "geographic areas where leading-edge anchor institutions and companies cluster and connect with start-ups, business incubators and accelerators. They are also physically compact, transit accessible, and technically-wired and offer mixed-use housing, office, and retail" (2014: 1). In other words, innovation districts bring together in one place the network of people, institutions, resources, and activities frequently cited as integral to the innovation process (Audretsch, 2003; Feldman, 1994; Malecki, 2010; Shearmur, Carrincazeaux, & Doloreux, 2016).

The aim of the innovation district, per Katz and Wagner, is to spur productive, sustainable, and inclusive development. Productive development is achieved by creating firms and jobs that assist in accelerating products to the market; sustainable development is achieved by focusing on density and compact growth; and inclusive development is achieved in two ways: by providing job opportunities across a diversity of positions for proximate low- and middle-income communities and by funding education programs that will enable residents to upgrade their skills and earn higher wages (Katz & Wagner, 2014: 1).

At their core, innovation districts are designed to exploit the benefits of agglomeration, e.g., the extra growth that is produced by clustering together firms, employees, and customers in the same geographic area in a manner that encourages frequent interaction (see Krugman, 1991; Porter, 1990). Proponents argue that four ingredients are critical for success. First, innovation districts must be place bound in order to maximize face-to-face interaction within a delimited space. Second, density is important. Design guidelines for innovation districts call for entertainment, retail, and housing amenities in close proximity to work; fiber optic cables to enable continuous public access to wireless connectivity; and the physical structures that support entrepreneurial activity, such as incubators and accelerators, research hospitals and universities, and legal and financial services. These design elements create an amenity-rich and walkable community that will be attractive to younger, high-skilled workers and the firms that employ them (Clark, Lloyd, Wong, & Jain, 2002; Florida, 2002; Lloyd, 2008). Third, networks are important to ensure that people are continuously interacting with each other and generating new ideas. Networks also create support for startups to scale. Despite the growth in remote working, face-to-face interactions remain an important aspect of innovation (Storper & Venables, 2004). Finally, though innovation districts are led and governed in many ways, often it is government representatives and prominent leaders from the surrounding anchor institutions (i.e., universities, hospitals, businesses, and foundations) who direct and fund the strategy through a public–private partnership.

Most innovation districts have been developed over post-industrial sites in older cities. Large urban economies are often touted as the main motors of economic growth (see for example Jessop, Brenner, & Jones, 2008), thus positioning the city as the best scale to concentrate activity (Glaeser, 2011; Pike, 2018). As with prior economic development strategies, the declaration of an innovation district serves as a branding mechanism to attract real estate development. Branding serves the purpose of rendering a place "safe" for investors (Klingmann, 2007) and also demonstrates an awareness of knowing the "right" elements needed to make a city a destination for tourism and investment. This succeeds in directing construction to places where investors and developers might have previously refrained from investing. Even when slated outside of the urban periphery, the declaration of an innovation district serves as an opportunity to build to the highest and best land use in accordance with market logics (Chapple et al., 2004; Dotzour, Grissom, Liu, & Pearson, 1990).

How Are Innovation Districts Working in Practice?

Do innovation districts really represent a better approach to urban redevelopment than previous strategies? For insight, we examine four cases of the concept in the United States: Cortex Innovation Community in St. Louis, MO; the Seaport Innovation District in Boston, MA; HUB RTP in North Carolina; and the Detroit Innovation District in Detroit, MI.

The Boston and North Carolina cases represent strong-market contexts. Strong-market cities have a robust entrepreneurial ecosystem. These are cities with research-based anchor universities, an abundance of talent in the form of skilled tech workers, and a pool of C-suite experts (Chief Executive Officer, Chief Financial Officer, Chief Technology Officer, etc.) who can advise emergent entrepreneurs. The Detroit and St. Louis cases represent weak-market contexts. Weak-market cities are declining, post-industrial cities that have struggled to transition economically and may lack some of the aforementioned resources, although they may have access to venture capital (Audirac, 2018; Beauregard, 2013; Mallach, Haase, & Kattori, 2017).

The cases also vary with respect to their metropolitan context. Three of these—Boston, Detroit, and St. Louis—are postindustrial sites in central cities. The fourth is a 100-acre innovation district within North Carolina's 7,000-acre Research Triangle Park, which is a suburban office park near the cities of Chapel Hill, Durham, and Raleigh (see Table 7.1).

St. Louis

The St. Louis Cortex Innovation Community, operated by the nonprofit Cortex organization, is a 200-acre campus for technology startups in the Midtown area of St. Louis, with a particular focus on biotechnology. Created in 2002 by five anchor institutions (Washington University in St. Louis, Saint Louis University, University of Missouri-St. Louis, BJC Healthcare, and the Missouri Botanical Garden), it initially started as a single building development that later spanned into

Table 7.1 Innovation district cases

Name	*Location*	*Founding*	*Acres*	*Governance Structure*
Cortex Innovation Community	St. Louis, Missouri, USA	2002	200	Managed by 501c3
Seaport Innovation District	Boston, Massachusetts, USA	2010	1,000	Managed by mayor
Hub RTP	Research Triangle Park, Raleigh-Durham North Carolina, USA	2012	100	Managed by 501c3
Detroit Innovation District	Detroit, Michigan, USA	2014	2,750	Advisory board representing public and private sector

a multi-building campus. It is the outcome of efforts dating to the 1980s to create a technology corridor between Washington University and Saint Louis University. Its founding partners (minus the Botanical Gardens) raised $29 million to fund Cortex initially, and in 2006 Cortex was granted the power of eminent domain within its proposed boundaries. However, significant legal proceedings were required to acquire the land.

In 2012, under new leadership, Cortex decided to rebrand as an innovation district in the legal documents it submitted to be designated as a TIF district (*St. Louis Innovation District Tax Increment Financing (TIF) Redevelopment Plan*, 2012). Cortex's decision to rebrand as an innovation district came on the heels of the Great Recession when little to no development was occurring in the area. Dennis Lower, then CEO, decided to transition the focus away from solely a bioresearch park and to create a live–work–play area with infrastructure more conducive to startup activity. Lower, who recognized the importance of mixed-use development to create a seamless flow between existing incubators and anchor institutions, applied for $158 million in TIF funding to address vacancy and blight in the area and to revitalize it through mixed-use retail, office, residential, and research developments, road and public infrastructure links, a healthcare facility, and open recreation space. Development of the site is ongoing, though it has already met many of its building targets. On account of demand, the Cortex Foundation is continuously updating the district master plan to expand beyond its boundaries (Feldt, 2018). Plant- and bio-sciences are no longer Cortex's only focus; smaller startup enterprises have flourished within the boundary of the innovation district, though these only represent a small percent of exits in the startup community with plant- and bio-sciences dominating venture capital funding (Smith, 2017).

Though Cortex has created jobs—as of 2020, the district hosted 425 companies employing 6,000 people—minority leaders have criticized Cortex for not meeting minority participation goals (Cortex Innovation Community, 2020). Minority hiring is an important goal because St. Louis is faced with a declining population, diminishing resources, large tracts of blighted land, and a racial divide (Gordon, 2008; Tighe & Ganning, 2015). The urban fabric of the Cortex Innovation Community and its surroundings is visual evidence of this divide, with boutique spaces sitting alongside structures that served a community faced with declining working-class opportunities such as a Goodwill retail store and outlet, the Salvation Army, Planned Parenthood, and Legal Services of Eastern Missouri, an organization dedicated to providing legal services to low-income communities. The ability for Cortex leadership to pay back the initial investment made by the area anchor institutions and to pay back TIF bonds is dependent on Cortex's financial success, which is based on increased capital investments and the rise of real estate values.

Boston

In January 2010, Boston's mayor Thomas Menino launched an initiative to redevelop a 1,000-acre swath of land into an urban laboratory of innovation and

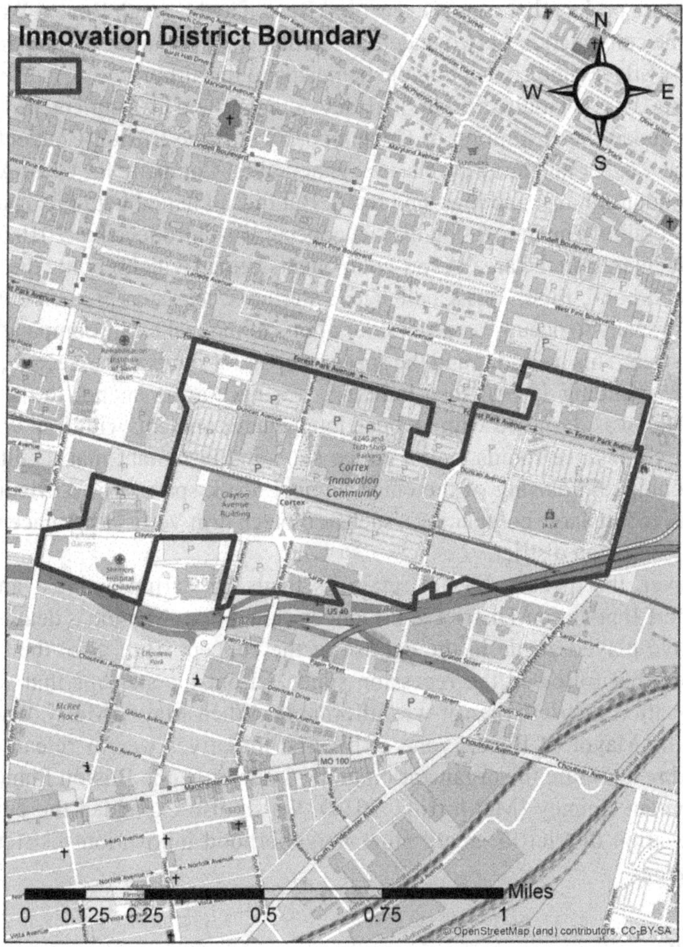

Figure 7.1 Cortex.

knowledge production. His strategy was to rebrand the four neighborhoods on the South Boston Waterfront peninsula as the Seaport Innovation District. Previously, in 1999, the Boston Redevelopment Authority (BRA) had hired urban design firm Copper, Robertson—the designers behind Australia's Sydney waterfront and Battery Park City in Manhattan—to develop a Seaport Public Realm Plan for the area. This was followed a year later by the South Boston Waterfront Municipal Harbor Plan. The Innovation District built on these prior plans.

The single largest factor Influencing development of the South Boston Waterfront was the Central Artery/Tunnel Project (1982–2006), a megaproject commonly referred to as the Big Dig. Believed to be the most expensive highway project in the United States ($14.6 billion, $2.6 billion over budget), the Big Dig was an infrastructure project that connected central Boston to the South Boston Waterfront through the construction of two underground tunnels (I-93 and I-90) and the extension of the Silver Line connecting the peninsula to the airport.

The Big Dig catapulted additional development and investment in the area, which included the Institution of Contemporary Art, the Boston Convention and Exhibition Center, and large purchases by the Fallon Company for Class A offices, luxury condominiums, and high-end retail. But this all halted with the onset of the Great Recession. The innovation district concept was introduced in 2010 to revive real estate interest in the area. A few components helped with branding the idea and the neighborhood. First, MassChallenge, a startup accelerator, partnered with The Fallon Company to occupy a vacant floor of one of the real estate company's buildings rent-free. This helped brand the space as open for innovators. Second, District Hall, a freestanding 12,000-sq ft building, was built as an anchor to the district. District Hall was designed with an open lounge environment, with free WIFI connectivity, conference rooms, and a coffee shop and restaurant that facilitated the early opening and closing of the entertainment–work space. Third, Menino used tax breaks to lure large companies—such as biotech firm Vertex Pharmaceuticals—into the innovation district.

However, the district did not retain its focus on startup activity, and today the Seaport Innovation District is no longer called that. Menino's death in 2014 might have played a role in the demise of the concept. But, also, the commercial boom that followed the recession squeezed out the startups that settled in the innovation district when it was affordable. When Mayor Marty Walsh took over as the Mayor of Boston, one of his first initiatives was to "clean out" the BRA and reconfigure the organization under a new title, the Boston Planning and Development Agency (McMorrow, 2014). Walsh also expanded the concept of the innovation district to create other neighborhood innovation districts in the Boston area.

Park Centre

Hub RTP is a proposed 100-acre innovation district at the center of a much larger 7,000-acre office park. Research Triangle Park (RTP) was founded in 1959 as a P3 with land acquisition by a private owner and public management (Abbott, 2018). By 1965 the non-profit Research Triangle Park Foundation (hereafter referred to as "the Foundation") took the helm as park manager, relying on income from selling and leasing land. RTP is credited for turning the Raleigh-Durham-Chapel Hill Triangle area into a thriving entrepreneurial ecosystem featuring an abundance of talent, venture capital, and C-Suite expertise. Today, RTP boasts of clustering firms in a wide array of sectors: ag tech, fin tech, bio pharma, clinical research, automotive, quantum computing, life sciences, clean tech, aerospace, cybersecurity, artificial intelligence, and machine learning. Hundreds of companies operate within the office park. The development of Hub RTP will push the research park in a new direction through the creation of an urbanized, transit-oriented, live–work–play district.

Figure 7.2 Boston.

At its inception, RTP was designed to span over 7,000 acres to prevent employees from competing firms within it from fraternizing with each other. The office park, which is located immediately east of Durham, is suburban in style. In addition to providing ample space for firms to develop their own campuses within RTP, zoning provision established an 8-acre minimum lot size, building set-backs of at least 150 feet from the road, and set-backs at least 100 feet from the side and back property lines (Rohe, 2012). Residential space was not included in the master plan and the park is designed for car commuting, not for transit.

In early 2014, the Foundation acquired 100 acres of land along I-40 for $17 million and designated the space as Park Center, later to be renamed Hub RTP. With Hub RTP, the previous siloed and expansive model of the RTP is supplemented by an innovation district to attract and retain entrepreneurs and young professionals.

The goal of Hub RTP is to accommodate 100,000 new jobs, build $2 billion worth of residential and retail amenities, and construct a rail path connecting Hub RTP to Raleigh, Durham, and Chapel Hill, the three surrounding cities that make up RTP. The site plan features an array of amenities to create a sense of "urbanity." These include street-level retail and entertainment, designated open recreation spaces, and housing in walkable proximity to work, all within a pedestrian and bicycle-friendly environment. The primary goal is to concentrate a density of people within the district to encourage interaction and spark innovation.

As of 2022, development at Hub RTP is ongoing. Office space is on sale, with two sites development-ready (housing between 150,000 and 300,000 sq ft of office space) and two buildings are occupant-ready (4000 Park Drive, with 73,000 sq ft of

Figure 7.3 Detroit.

Class A office space available, and Frontier RTP, a refurbished 1980s office building with 500,000 sq ft). Retail space is also slotted for development, with a project delivery date of late 2023. Because of the growth in the surrounding area, Hub RTP is, in a sense, building the city-core to tie together more remote developments. It is branded as the center of the largest tech park in the United States ("Location," n.d.). Unlike other proposed innovation districts, Hub RTP is not a postindustrial urban site. Hub RTP sits on land already managed—through previously not owned—by the Foundation. And though most of Hub RTP is greenfield development, RTP's newest development demonstrates that innovation districts are not uniquely advantageous to cities and that there will be suburban competition.

Detroit

The Detroit Innovation District (DiD) was declared in the summer of 2014, but it built on decades of effort to redevelop the greater downtown through various placemaking initiatives. Like other cities, Detroit adopted numerous economic development and tourist attraction schemes, including waterfront redevelopment, casino construction, and sport-led regeneration (Eisenschitz, 2010; Grodach & Loukaitou-Sideris, 2007; Klingmann, 2007).

The proposed Detroit Innovation District spanned 4.3 square miles and covered much of Greater Downtown, including the Central Business District, Midtown, and Corktown. Anchor institutions in the district include Wayne State University, two major hospital systems, an established business incubator, and the College of Creative Studies. Michigan Governor Rick Snyder declared the district and then rolled over responsibility to the city. Mayor Mike Duggan supported the initiative with a public announcement, but foundation leaders—especially those from Hudson-Webber's New Economy Initiative (NEI)—led the charge. The district also had an 18-person advisory committee that steered its development. Members included representatives from Wayne State University's entrepreneurship hub, TechTown; Henry Ford Health System; General Motors; business owner, venture capitalist, and real estate mogul Dan Gilbert's series of companies; non-profit economic development organizations, such as Midtown Inc. and Invest Detroit; three area universities (Wayne State University, University of Michigan, and Michigan State University); and some area entrepreneurs. The focus of the advisory board was to target 1) Physical Place, 2) Innovation and Commercialization, and 3) Building Detroit's Knowledge Economy. Subgroups devoted to these goals, made up of ten to twelve people, were tasked with presenting concrete plans for their respective areas to the advisory committee.

The board produced several reports, but the district never took concrete shape. By 2016, initial funding had run out and the leaders were no longer meeting on a regular basis. New funds were raised, but the efforts for the DiD were hampered by leadership conflicts, including a disagreement between representatives of the Detroit Economic Growth Corporation (DEGC) and the New Economy Initiative leadership. By 2017, DEGC had opted to move toward a high-tech, high-growth business development cluster strategy targeting the food sector, medical health, and mobility, opposite NEI's intention for a broader, more inclusive implementation that benefits Detroit's low-income population more directly.

The DiD differs from the previous cases in that it did not focus on a specific geography; the land was not owned through a TIF or a non-profit foundation; and it did not have a master plan guiding its development. Though the previous iteration of the DiD has disbanded, more focused attempts to create innovation districts have surfaced. In July 2022, the University of Michigan received a $100 million appropriation from the state legislature for the development of the Detroit Innovation Center, a collaboration with real estate firms Related Companies and Olympia Development of Michigan, which will include a new graduate school campus downtown focused on innovation (Eggert, 2022). The State of Michigan, the City

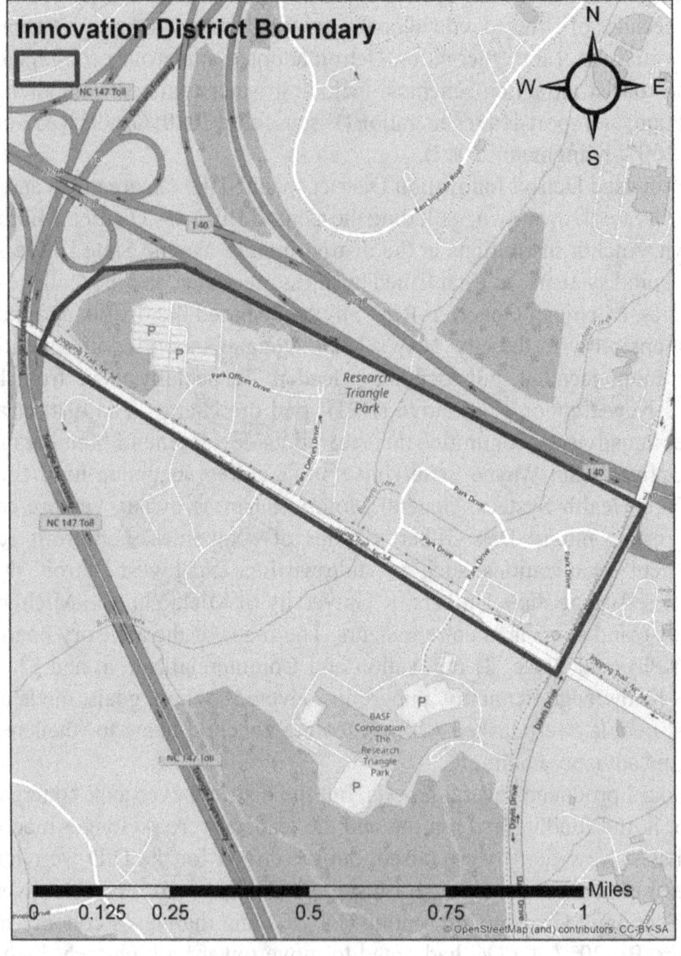

Figure 7.4 Park Centre.

of Detroit, and Ford have also agreed to create a "Mobility Innovation District" around the renovated Michigan Central Station in Corktown (Michigan Central, 2022). While the previous innovation district in Detroit could ultimately be classified as a short-term re-branding exercise for Downtown and Midtown, these new initiatives involve the master-planned redevelopment of specific neighborhoods, like the other cases presented here.

Do Innovation Districts Live Up to the Hype?

The cases above raise a series of questions that we aim to address here.

First, have innovation districts succeeded in their mission of supporting startup companies and innovation? Cortex in St. Louis has succeeded in creating jobs in the high-tech sector, and the Boston Innovation District also demonstrated growth in the high-tech sector following the Great Recession. However, Cortex is running out of space to accommodate new startups, and in Boston many startups have been priced out, with much of the startup activity relocating to the vacant buildings that the legal and financial firms are leaving behind as they move their offices to the high-end Seaport Innovation District (Martin, 2016; McMorrow, 2012). In Detroit, where the innovation district concept was not implemented, it remains to be seen whether two new efforts to spark startup activity will be successful, although the central business district has already seen an influx of startups connected to Dan Gilbert's family of companies. Likewise, with Hub RTP, it is too early to predict the district's outcomes.

A second question that arises from the cases is whether innovation districts have succeeded in their mission to boost high-density, mixed-use, live–work–play urbanism. While Cortex expanded from a single building to an entire district with a range of office buildings, cafes, restaurants, and open spaces for outdoor conferences and food trucks, the lack of housing within the innovation district prevents it from becoming a fully live–work–play location. Boston's innovation district does include housing, but the district prioritized luxury developments (as of June 2022 the median home price was listed as $1.6 million ("Seaport District, Boston, MA," 2022)) thus preventing anyone but the wealthy from living there. Detroit is experiencing increased mixed-use investment within its Greater Downtown, but this trend is not attributable to the planning for an innovation district but rather to independent investments by the various parties to the planning effort, e.g., the expansion of Dan Gilbert's family of companies, Ford's recent investments in the Corktown neighborhood, and investments by Wayne State University. Hub RTP is still in development, so it is not yet clear whether it will successfully incorporate housing and transit into what remains principally a suburban office park.

Finally, do innovation districts merit public subsidy? As one of the newer economic development tools, innovation districts could show promise for both the real estate developers who build them and the communities that host them. Because most innovation districts are built in central cities—often over abandoned industrial sites—real estate developers who participate in these projects can potentially leverage a wide variety of real estate incentives. Developments could qualify for municipal and state-level incentives for brownfield redevelopment, historic preservation, affordable housing development, job creation, and more.[1] The federal government also provides a series of incentive programs that could be leveraged by real estate developers to build innovation districts, including historic preservation tax credits and Opportunity Zones, which are a newer incentive generated from the Tax Cut and Jobs Act of 2017 (Gramlich, 2022). Energy efficiency and health incentives could also be another source to offset higher construction costs and provide upfront capital.

But do innovation districts deserve these incentives? When used appropriately, economic development incentive programs can produce a positive ripple effect on a community's development, thus generating additional revenues through expanding the tax base, increasing property values, adding amenities, creating job opportunities, attracting talent, and increasing consumer spending. However, there are reasons to doubt that innovation districts will fulfill these goals. Though empirical data remains limited, innovation districts have already demonstrated some negative impacts on surrounding communities in terms of polarized division of labor, housing unaffordability, and income inequality (Zandiatashbar & Kayanan, 2020). Nor are there any clear examples of innovation districts that have met inclusive development objectives (Kayanan, Drucker, & Renski, 2022). So even if they do succeed in creating service level jobs, the idea of a live–work–play environment will not equally benefit all. In fact, it may increase commuting distances for the individuals who can no longer afford to live in the district. Therefore, any incentives used to offset construction costs must keep in mind the importance of affordable housing, the diversity of amenities, and the provision of schools and community services that can make the space of the innovation district inclusive.

Conclusion

This chapter explained the innovation district concept and then critically examined its implementation as an effective tool for inclusive economic and community development in four locations in the United States. Furthermore, it contextualized the rise of innovation districts, comparing them with previous economic development tools used to revitalize neighborhoods and transition cities from an industrial to a post-industrial economic base. Core to the argument throughout the chapter is that there are risks to place-based development tools like BIDs, TIFs, and EZs: the funneling of resources into some neighborhoods to the neglect of others; ceding control over public space to unelected business groups; the financial risk that comes if property values do not rise; the displacement pressures that come if property values do rise significantly, causing gentrification; and the hoarding of revenue to further boost downtown, with limited cross-subsidization or benefit to non-targeted areas. These downsides can feed into a broader criticism of post-industrial urban redevelopment. Over time, these tools contribute to a less equal city, with a greater share of individuals stuck in poorly paid service positions watching development disparity increase between the haves and the have nots.

Economic developers and growth machine coalitions willingly embrace innovation district strategy because of the opportunities it presents to transition into a tech economy, create jobs, promote vibrancy, and, through spillovers, generate regional wealth. Whether or not innovation district strategy will meet these objectives remains to be seen. In Boston and St. Louis, innovation districts have contributed to job growth and rising property values. In Detroit and Research Triangle Park, innovation districts are still in development. Going forward, the question is whether and how this new strategy, like the EZs, BIDs, and TIFs that came before, can influence private sector investment for the benefit of the community.

Note

1 Because each state and municipality differ in terms of their incentives and their stipulations, it is important to closely research the programs and to determine which ones apply best for each locality.

8 Industrial Development and Public–Private Partnership

The Enigma Case Study

Brockton Hall

Introduction

Public–private partnership is a key role in industrial development in many parts of the United States. It can be both adversarial and mutually beneficial, depending on the position of the private developer and the community in which the project is built. This can present itself in many different ways, but the three most common touchpoints are during the entitlement phase, road and utility improvements, and grants and tax abatements.

On the adversarial side, many communities with good logistics connectivity and infrastructure struggle with the balance between lower income residential areas, interplay of commercial and heavy truck traffic with residential, and the demand for industrial development in their communities. An example would be logistics development in Southern California and the Inland Empire. Developers have to work closely with the local community and various special interests groups to secure the necessary approvals to build these buildings. These can include light and sound regulations and caps, height and density restrictions, offsite improvements of roads and ingress and egress points, as well as sustainable building practices in some cases. Even in more amenable communities, being able to work with the community to develop a feasible plan and work through entitlements is a crucial part of the developers' job. Once completed, these higher barriers to entry markets can yield significantly higher returns because of the difficulty and time to entitle compared against the extremely high demand for space within these markets.

Industrial developers also go in the opposite direction and look for pro-industry, pro-economic development communities which are looking to incentivize industrial development to create jobs and tax bases for their communities. These can be higher risk but also lower barriers to entry with simpler entitlement processes, lower land and development costs, and a more favorable viewing public citizenry. This process is common in many former textile communities in the Sunbelt and rustbelt cities in the Midwest. Its also very common in rural communities with little industry which are looking to create jobs for their citizens. Developers which can develop good relationships with these communities can use infrastructure grants, faster development timelines, and public tax abatements to drive down development costs and increase potential returns.

DOI: 10.1201/9781003222934-10

In both cases, there are very few industrial projects built without a high degree of interaction between developer and public officials, making a good knowledge of P3 mechanisms and the entitlement process crucial for successful projects.

The Case Study: Enigma's P3 Adventure[1]

Background

The story begins with a national development company which was looking for a site to develop large-scale warehouse and distribution facilities on a speculative basis in the area along the I-85 belt in the Carolina's. The developer, Enigma Development Group, had identified the submarket in question to develop these speculative facilities due to the increased demand for regional distribution centers and manufacturing plants in the market in question. The goal was to develop "Big-Box" warehouses from 500,000 to 1,000,000+ square feet, which were cross dock in orientation to meet the needs of these large requirements in the region. At the time this was one of the first projects of this nature which would be undertaken in the market and would represent the largest buildings ever developed on a speculative basis.

There were several factors which needed to be considered during the site selection process. The first issue was finding a site which was in one of the primary submarkets in the market with developable topography, clear from any major geotechnical and wetland issues, close to major interstates (namely I-85), close to an intermodal facility or rail yard, LTL, and small parcel service and with all major utilities on site. One of the other major issues was checking all of those boxes while also finding a site that could accommodate the footprint of these large-scale buildings, which can routinely be larger than 30 acres under roof, and the limits of disturbance can be over 1,000 feet wide and 2,500 feet long. There is also a lot of grading that takes place to mass grade these sites and cut/fills can be as much as 100 feet in some areas of development. A prototypical "big box" distribution building is 1,100,000 square feet at a dimension of 620 feet wide and 1,700 feet long. These buildings are 36–40 feet in clear height, have truck courts and dock doors on either side, with dock doors and opposing trailer storage typically 185 feet wide. Once you consider the grading that must take place to get to a consistent building elevation, you end up needing over 100 acres to accommodate these buildings.

All of this to say that these are very large buildings with very big footprints, and when you are buying land by the acre every acre of coverage counts. Hopefully this gives you a better idea of the size of these developments and the land needed to accommodate them. The challenge for Enigma Development was to be able to check all those boxes while maximizing coverage and playing the world's largest game of Tetris to fit these buildings on the site while not impacting wetlands or blue line streams.

Enigma toured the market and met with several local brokers and community officials to evaluate potential sites and get to know the government authorities which would have jurisdiction over the development. The Enigma Development team identified the Highway 211 submarket as the preferred submarket to build spec. So why was this submarket selected? 1) It was located equidistant from the two major population centers in the market so it had good access to labor within a 45-minute drive time; 2) it was already an existing industrialized area with most of the national LTL freight carriers in addition to shipping terminals for FedEx and UPS; 3) it was close to the international airport for the market with air freight capacity and a FedEx Air Freight terminal; but, most importantly, 4) it was the submarket which contained a major Norfolk Southern Intermodal Terminal called Riverside Terminal which brought sea containers of cargo from the coastal ports inland to the market.

What is an intermodal terminal, you ask? Intermodal in Supply Chain verbiage means a node or logistics point where freight changes Mode of Transportation, hence "intermodal" Freight transitions from rail to truck, truck to rail, or ship to rail and rail to ship. In today's world of online shopping and high-speed logistics, an intermodal terminal can add a lot of value to a company that has to import bulk commodities or finished goods from overseas and then stage it in a distribution center. How does it add value? Answer: it allows a company like Target or Walmart to receive goods from its vendors in Europe and Asia at a coastal port from a container ship and move those goods in bulk by rail to a location further inland, preferably on a major interstate with logistics significance with the widest footprint of consumers possible. The reason for using rail is that it's the cheapest way to move bulk containers from Point A to Point B. For those interested in Environmental Social Governance (ESG), it also significantly reduces the amount of CO2 which is released into the atmosphere, and truck traffic on roads, because it takes containers off the interstate which would otherwise be trucked to their destination from the port. The alternative is to put the warehouse close to the port terminal to receive the shipping containers. This is typically more expensive than inland markets, more difficult to develop due to wetlands, and most importantly half the area you are trying to ship to is under water.

The Site

Now back to Enigma and their site selection. Considering that we have checked our macro locational boxes to find the right submarket to build in, it's time to find a site. The Highway 211 submarket was an up-and-coming submarket which had a lot of good sites but had previously been overlooked by developers because of a lack of utilities and adequate infrastructure. Most of the development had been skewed to the next two interstate exits where there was existing sewer infrastructure and more established industrial parks. The submarket was defined by Highway 211, which was a five-lane highway that both ran North and South and perpendicular with I-85. The Riverside Terminal was to the extreme North of the submarket along Highway 211 between the City of Eureka to the west and the Town of Gantt to the east, which were the two closest municipalities to the submarket. The submarket as a whole is located in Hall County, which would be the statutory authority for any real estate tax-based incentives that could be garnered by the project.

Enigma was looking ideally for a site where they could build three to five buildings between 500,000 and 1,000,000 square feet, with a particular interest in any site which could accommodate a building of 1.5 million square feet or larger as these would be the most unique and therefore the most valuable for import centers looking to bring in containers from the intermodal terminal. In order to accomplish this, they would need a site of roughly 200–300 acres or multiple sites within close proximity to one another. Unfortunately, the submarket was mostly farm and pastureland and had seen very little investment into roads and sewer infrastructure. There were also several significant blue line streams[2] that ran throughout the submarket, which divided and impacted many of the larger tracts that would otherwise have been developable.

Luckily for Enigma, there was one site which had recently been listed for sale by a local investor who had owned the property for several years. His name was Mr. McDoogle and he had affectionally named the site McDoogle Park after his namesake. He had also assembled the adjoining residential properties with the hope of one day selling the entire assemblage as one unincumbered tract of land. Mr. McDoogle accomplished this by buying some of the tracts over the years and purchasing Sale Options on the remaining parcels. The assemblage totaled 225 acres and was priced at $65,000 per acre, which at the time was significantly higher than any other land price in the market for a bulk land sale. Although the price was higher than anticipated, it met many of Enigma's requirements. There were two streams running north and south which spilt the site into three pieces, but the central piece was about 60 percent of the site acreage and was unincumbered by wetlands.

When evaluating the center section of the site, Enigma estimated that it could fit in several buildings, or one large building of over 1.5 million square feet, without impacting the wetlands. They also determined that they could build some additional smaller buildings on the remainder sections of the site or sell less desirable acreage to single family developers who could more easily work around those impediments. The site was bordered by two-lane secondary roads, which were in disrepair and in need of improvement, but luckily the main entry road for the 1.5 million-square-foot box was about to be improved as part of the incentive package for a large automotive manufacturer across the street. It also had access to a municipal sewer via a pump station south of the site owned and operated by the City of Eureka. The Gas and Water lines adjacent to the property were also owned by the City of Eureka and located in the public road rights of way. It was also located centrally between the Riverside Terminal and I-85, three miles away from each. The site was gently rolling pastureland and was generally clear of trees except along the wetlands. Mr. McDoogle leased the property to a cattle farmer who was the sole occupant other than the residences along the perimeter of the property. After doing their due diligence on the area, Enigma determined that this was the best available site in the area for its development. The next step was to tie up the site and start its due diligence and pre-development activities.

Obtaining the Property: McDoogle Farms

During this process, Enigma was assisted by a local brokerage firm, Best Industrial Brokerage, which assisted in the site selection process and served as "boots on the ground" for Enigma's development team which was located out of market. The brokerage team's local relationships and market expertise were essential in the identification of the submarket and ultimately the selection of the site in question. They also had a good understanding of the local political groups and utility authorities who would serve the property or have jurisdiction over any project that took place in the area. By a happy coincidence, Best Industrial also had a good relationship with Mr. McDoogle and his attorney Mr. Smith, who assisted Mr. McDoogle in managing his estate and represented him in all his real estate transactions.

Once Enigma determined that McDoogle Park was the location of choice, it engaged with McDoogle's attorney to negotiate the purchase of the property. There were a couple of items which directed the discussions with regards to business terms. First, the property would need to be annexed into the City of Eureka in order to get access to its utility infrastructure. The property was currently in the unincorporated Hall County that did not have any utilities in the area, which was a big reason why this submarket was historically undeveloped. The City of Eureka had the highest millage in the area and most developers tended to choose to develop in the unincorporated county elsewhere in the market for that reason. They would need to have enough time to conduct due diligence as well as go through a three-month annexation process, so they needed enough time from McDoogle to complete these efforts. Second, Enigma was very sensitive to remaining competitive in the market in terms of pricing. There were several large Industrial Parks and Land Tracts in the market owned by non-developer entities. These private entities were local land developers and private funds, which could sell their property to users directly or partner with build to suit developers with tenants in tow. At the time these were lower in prices than what Mr. McDoogle was marketing the site for sale. They needed to make sure that they had a financially sound development plan and could offer their speculative building at competitive lease rates when compared with these sites. This, coupled with the fact that they would need to overcome a higher tax burden by being incorporated into the City of Eureka, meant that they needed to get some relief from Mr. McDoogle on the land price, which was $20,000 per acre over the next closest comparable land price. Lastly, Enigma knew that it would be a significant amount of effort to work out all the necessary agreements with the City along with working through the site and infrastructure issues in order to make the development successful. The roads needed to be upgraded and the utilities were undersized for the use.

The principals at Enigma also agreed that, if they were going to make this a viable industrial subsector of the Highway 211 submarket, they wanted to maximize their investment. If Enigma was going to do all that work and front end the infrastructure, they wanted to incorporate as much property into the development as possible for future phases of the park. That way they could capitalize on their initial success by controlling a larger piece of the submarket. With all this in mind, Enigma would need time so they could work out all the issues, do due diligence, tie up adjoining properties, and they needed to get some relief on the price.

Best Industrial approached Mr. McDoogle's attorney, Mr. Smith, about the initiative Enigma had in mind and explained some of their concerns about the entitlements and the need for a price reduction based on the size of the tract. Mr. McDoogle had reason to expect a higher price for his property. Adjacent to McDoogle Park, there had been a land sale at the desired price point, but it was about a third of the size and was purchased by a manufacturing company. Generally, land sellers in the areas could garner a higher price by selling to users because of their desire to be in the submarket close to the intermodal terminal, it being a relatively low input into their long-term capital investment in their projects, and not having the need to garner some type of return on the development. Real estate is after all less than 5 percent of a typical industrial user's cost to operate a facility. Long story short, users typically pay a higher price than developers if they want the site. Mr. McDoogle also had another tract in the area listed at the same price about a quarter of a mile away from McDoogle Park, which had seen a lot of interest from developers as well. McDoogle Park had been on the market for several years at this point and had been passed over for many projects both by developers and users due to the development issues and largely the sale price.

Ultimately, Enigma negotiated with Mr. Smith and finally agreed to purchase the property for $40,000 per acre and agreed to purchase his additional tract a quarter of a mile away at full price value to help make the deal more attractive for Mr. McDoogle. He also agreed to 90 days of initial due diligence with the ability to extend the due diligence for additional non-refundable earnest money deposits by two additional 30-day periods. Thus, checking all their boxes, timing, price reduction, and buying more property for future development. Taking this a step further, they negotiated with a local tree farmer, Mr. Green, who owned a nursery that adjoined the northern property line. This additional 84-acre tract allowed Enigma to make land on the eastern side of the park, which had had limited utility, into a much larger footprint site which could accommodate an additional 1 million-square-foot building along with adding an additional development site for roughly 500,000 square feet. At this point Enigma had assembled and put under contract more than ten parcels of land to amalgamate a roughly 325-acre park with an additional 40-acre site located down the road for future use.

The Public–Private Partnership Component

Now let us consider the public–private partnership piece of our story. As mentioned before, most of the development to date had taken place in unincorporated areas of Hall County for four major reasons: 1) lower land prices; 2) lower taxes; 3) pro-business outlook; and 4) existing sewer infrastructure.

Enigma was convinced that the emergence of the Riverside Terminal in the Highway 211 submarket and recent scarcity of developable sites was going to push development and prospective tenants to the Highway 211 submarket. Spoiler: they were right, but that's a story for another day. Their main challenge was that they would either need to put more money into the development budget to pay for these infrastructure improvements, which they didn't have, or get some public side assistance to help repair and improve the surrounding infrastructure. These improvements were badly needed in the area as most of the secondary and tertiary roads were farm roads with narrow shoulders and not designed for the truck traffic they would inevitably see over the coming years. There were also very few traffic mitigation measures such as traffic lights and turn lanes. These would all need to be upgraded to handle heavier traffic. The two main roads servicing the property were Bricktown Road, which was a two-lane road on the southern border of the property, and Kelly Mill Road, which was also a two-lane road on the eastern border of the site. These both had limited rights of way available and had limited shoulders. Enigma also had to annex into the City of Eureka to get access to sewers, as they were the only provider in the area. Several years prior, the City of Eureka, which had a sewer treatment station in the submarket, extended a small force main line to a pump station to the bottom of the creek basin. This was done to facilitate a cluster of manufacturing projects including the user which Mr. McDoogle used to base his sale price. They positioned it in such a way that it could benefit other industrial customers moving to the area in the future. Other than this small sewer line, there was no other significant sewer infrastructure in the area and all the other warehouses in the area were served by on-site septic systems.

The development of this sewer infrastructure wasn't a purely altruistic play by the City of Eureka. The City used its utility services to spot annex any servicing property in order to grow its tax base, which had been a very successful strategy. To receive a "will serve" letter and service agreement, landowners would have annexed their property into the City. Due to the City's investment, they were the only sewer provider in the area. The only other provider close was the Town of Ganntt.

Annexation also allowed the City of Eureka to serve power to the annexed property, along with natural gas and water. Natural gas and water were the chief concerns but the potential for being restricted from using a large private utility or cooperatives could put Enigma at a significant competitive disadvantage when competing with sites with lower taxes and a regional electrical utility.

Enigma also needed a clearer interpretation or exemption from the City's business license fees because the City Code didn't adequately define how gross sales were determined for distribution facilities. They were worried that it would be perceived by a fulfillment user, actual or not, as a major liability and business risk. In concept, if Amazon leased one of the buildings and had to apply for a business license in the City of Eureka, it could be interpreted under the City Code that the business license fees would be calculated based on all the online sales that take place in that building. This could create astronomical business license fees for an e-commerce retailer like Amazon, which could poison the market against the development. For these reasons, Enigma wanted to work with Hall County and the City of Eureka to get some assistance in regards to: 1) funding or contributing towards offsite costs made up from upgrading public infrastructure; 2) getting some abatements towards real estate taxes to make it competitive with surrounding sites and communities; 3) waiving or capping the business license fee liability; and 4) getting an agreement that, in the event that the local power utility could not meet the needs of a potential customer, Enigma had the right to seek an alternative remedy.

Keep in mind, it wasn't without benefit to both Hall County and the City of Eureka for this development to take place. The majority of the additional millage wouldn't have gone to benefit the development and included public sanitation, police and emergency services, and city capital improvements, none of which the development would use because of its tertiary nature to the City. Due to the size of the industrial users, they would contract directly for sanitation and security services. The development contemplated would also constitute over 3.5 million square feet of development valued at over $350 million dollars at full build-out and could create at a minimum 300–500 good paying jobs and potentially thousands of jobs if the site was dedicated to a large manufacturer. Once put into service, a single building could generate over $1.5 million in annual real estate tax revenue and those taxes are perpetual.

From Enigma's point of view these public improvements, jobs, and taxes would all be beneficial to the community and would leave a long-lasting impact on the tax revenue for both City and County. What they were asking for in return were short-term concessions and cost sharing for road and utility upgrades, which the community admitted had fallen below minimum standards. Together, the Enigma and Best Industrial Teams would need to figure out if these goals would be achievable and the best way to execute them if so. Luckily for Enigma, the County had been very successful in recruiting new companies to the area and had a very pro-business mindset. The Best Industrial team had had a lot of success working with Hall County through several different programs and private grants and had tackled similar problems on other projects.

Best Industrial recommended three local law firms which each had practices relating to economic development and economic development incentive negotiations. These firms represented incoming industrial and office users in the negotiation and securing of tax and business incentives used to recruit users, creating jobs and developing projects in the local communities. In some cases, these firms also served as outside counsel for these communities and assisted them in the crafting of legislation and ordinances regarding these grants and incentives. They were basically the hired guns to get the best deal possible and back-end administrators for compliance during the term of these agreements. It was a lucrative business line for these firms, albeit a very niche practice, with a small group of lawyers who were considered experts in the area. Enigma interviewed all groups and ultimately selected Briggs, Langston, and Associates (BLA), who had done the most work in the City of Eureka and had recently secured entitlements on a highly contested hotel project in the City.

Outside of matching funds and cash grants, which typically were only provided as closing funds for economic development projects, there were a couple of major programs which BLA introduced to Enigma. These were the PILOT program, SSRC program, and Utility Tax Credit (UTC) program. PILOT most people are familiar with; it stands for Payment in Lieu of Tax, also sometimes called Fee in Lieu of Tax (FILOT), and varies state to state based on their state statute. In this case the PILOT allowed the County to lock in an agreed upon assessed value, usually with pricing adjustments built in for 10–30 years. The benefit to this was that it locked in a negotiated payment in lieu of a tax payment, which could be negotiated in the user's, or in this case the developer's, favor and not be subject to periodic reassessment or special assessment. Also important is that it exempted the property from spot reassessments based on the future sales of the property.

The SSRC stands for Special Source Revenue Credit and would allow for a temporary reduction of ad valorem taxes levied on the subject property based on an investment by the user into public or dedicated infrastructure. For example, if ABC Auto Plant invested $1,000,000 into a sewer line to serve their factory and the neighboring properties and then agreed to dedicate it back to the County, ABC would be reimbursed for 80 percent of the value, or $800,000 over five years, in the form of a real estate tax credit during the repayment period. This schedule is based on state statute and can be as little as 50 percent to as much as 80 percent of the value of the infrastructure invested and could be credited over five to seven years.

Both the SSRC and the PILOT programs were regulated by the state and implemented at County level. These are also some of the most impactful to the development long term. Both programs required three readings by county council to pass. The last program is a simple but effective one called the Utility Tax Credit program. This allows private utilities to offset their corporate state income tax with a credit up to a percentage of its obligation by investing money in public infrastructure valued at the same amount of the tax liability in question. This creates an incentive for private utilities to invest in roads, bridges, water, and sewer lines in their service territories which in term are, or would be, customers of their utilities. City, County, and other public entities can't use this program because they are already exempt from state income taxes.

BLA started conversations at the County level as they were the most amenable to the development in question. Due to its size and population, Hall County was most interested in the project because of the potential to create jobs in the community, which continued to see a high percentage of in-migration from around the United States. The County was also responsible for the maintenance of the roads in the submarket. For these reasons, it was very supportive of the project as it would front end the cost of upgrades that it was already planning and create new jobs. Luckily, the County had already agreed to mill and resurface Bricktown Road as part of the incentive package for a $300 million automotive plant, which was under construction south of the development on the opposite side of Bricktown

Road from the project. After discussions with BLA, the County agreed to providing a PILOT for the project to lock in the tax rate and assessments but wanted to understand the full scope of infrastructure improvements and costs before committing to further incentives. Some of the infrastructure discussed included improving Kelly Mill Road and contributing a Right of Way, as well as participating in the improvements of Bricktown Road such that it would be widened, and a turn lane and median would be installed.

The next task was working with the City of Eureka to determine what assistance, if any, would be provided. All the other utilities serving the project (gas, sewer, water, power) would be provided by the City of Eureka's City Public Works (CPW). The CPW water line would need to be upgraded to reach pressures and flows high enough to support an ESFR (Early Suppression Fast Response) sprinkler system or the project would additionally be burdened by fire suppression water tanks for each building. The gas was adequate and there were CPW power lines close enough to serve the site. There was a lift station for a sewer across Bricktown Road from the project, but it would additionally need to be upgraded to serve the project.

The City of Eureka agreed to upgrade these utilities to serve the project, allowing a carve out for Enigma to use a larger electric utility in the event of an inability to serve, the stipulation being that the property would have to be annexed into the City of Eureka for service. This was expected but, as mentioned previously, the City of Eureka had the highest tax burden in the area, so Enigma's first request was a short-term, or partial, abatement on the additional millage over the average county millage in the area. This was mainly for recruitment reasons during the lease-up process. The City of Eureka was not open to any tax reduction: strike one.

The next goal was getting a more defined explanation of the business license fees and a carve out for any e-commerce users which may occupy the park. Enigma used the Amazon example given previously in the chapter, but the City disagreed with the assessment and refused to make an exception or special provision for McDoogle Park. Although BLA agreed that it was a legal gray area, which could potentially create exposure for future e-commerce tenants in the park, the City disagreed: strike two. The City also did not agree to provide any UTC funding for road improvements through its utilities: strike three. With the upgrades to the utilities, Enigma could make the numbers work from an infrastructure standpoint, assuming it received some assistance from the County for road improvements, but the stance on the business license fees and City taxes was concerning. This was a large investment by Enigma in an unproven area of the market, and it would be developing the largest buildings ever built in the market. There was active build to suit activity in the area of a similar size and scale but any actual or perceived disadvantage when recruiting these "big box" users substantially increased the risk and investment committee scrutiny.

Seeing that the conversations with the City of Eureka weren't making a lot of progress, Best Industrial on Enigma's bequest started working with BLA to determine what alternatives, if any, were available in the area for utility service. There were other providers in the County which had lines servicing the area with water, gas, and power but there wasn't another sewer provider in the area other than CPW. All the other buildings in the submarket not served by CPW were on private septic tank systems. There was also a gentleman's agreement that other sewer providers wouldn't cross west of Kelly Mill Road to serve users there as the Town of Ganntt and Hall County Sanitary Sewer District both had service further South in the 221 Submarket as well as the US-1 submarket which was the next exit to the east. The Town of Ganntt was increasingly frustrated with the City of Eureka to the west, as it was adversarial to new development in the surrounding area which would benefit both the Town of Ganntt and the City of Eureka due to their close proximity to each other. Based on the guidance of a local official, Best Industrial and Enigma met with the Mayor, Town Administrator, and Fire Chief of the Town of Ganntt to discuss the project and their vision for the area. The Town was very open to working with Enigma to help develop the project and viewed this as an opportunity to extend utilities to undeveloped areas between the McDoogle Park and the Town limits to serve future development in the Town's service territory.

The Town of Ganntt was much smaller than the City of Eureka in size and annual revenue and had been particularly affected by a series of mill closures. They didn't have the capital or potential for the bond raise necessary to pay for speculative sewer extensions of this nature. However, one thing the Town did have was a tremendous amount of sewer treatment infrastructure because one of its shutdown mills had an independent treatment facility, with the ability to handle industrial effluence, left over from the operations of the old mill, which was still active. The Town was using less than 50 percent of the capacity of the plant at the time of development, certainly more than Enigma could ever use. They also had a 12" water main along Kelly Mill Road, which was much larger than CPW's line and served other users in the area. Power could be provided by large regional providers with lines close to the project which could serve the buildings at no additional cost. The upside for the Town was that they could trade near term tax revenue that they didn't have access to, regardless, for a developer to front end the cost of a sewer main to serve the basin around McDoogle Park. The Town would get the revenue once the incentives burned off and it would open up additional land sites for industrial development which the Town wouldn't have to offer additional incentives to. Based on these factors, the Town agreed that it would extend a sewer line from its nearest sewer substation, roughly two miles away; part of gaining the rights to service the site also involved annexation. Due to the increased millage associated with the annexation into the Town, the Town agreed to rebate 90 percent of the real estate taxes for the first five years of the development. All other required utilities already existed in the immediate vicinity of the project. The only challenge with this strategy to annex into Ganntt was that the sewer line would need to be extended a much longer distance and would require a right of way acquisition in order to secure the route to the project cross country.

While Best Industrial worked with the Town to confirm the route and discuss potential right of way acquisition terms with the property owners, Engima turned their attention to the increased cost of the sewer extension and potential solutions for incorporating it into the development. Enigma revisited their pro forma and determined that the additional cost of the sewer in addition to the required road improvements would cause a situation whereby they would either need to reduce their development yield on the project or increase the lease rate for the buildings. They couldn't reduce the development yield for the project, and increasing the lease rate could put them at a disadvantage versus built-to-suit options and future speculative competition.

There was one option that was still on the table which wouldn't help their cost problem but would reduce the overall gross rate of the project. That option was to discuss a special source revenue credit (SSRC) from Hall County. If they were able to secure an SSRC agreement from the County, it would reduce Enigma's real estate taxes based on the value of the public infrastructure that Enigma would install over a period of time.

With this in mind, Enigma, with BLA in tow, set up a meeting to talk with the economic development officials from Hall County. Enigma laid out their case to Hall County, specifically that the project could not move forward without the extension of the sewer line from the Town of Ganntt and road improvements along Bricktown Road. Hall County was open to the idea but pressed that the only reason the sewer extension was needed was because they weren't annexing into the City of Eureka for utility service. Enigma countered that, based on the current terms with the City of Eureka, it was a non-starter and that they couldn't take the risk of a failure to stabilize the buildings because of an above-market gross rate or a tenant rejection of the City licensing fees. The meeting ended with an SSRC on the table pending further discussions with the City of Eureka.

While Enigma had been in discussions with the County, Best Industrial was working with Mr. Bruce, a local tree farm owner whose property was between the development and the Town of Ganntt's sewer line. Luckily, the Town had a relationship with Mr. Bruce and worked with the Best Industrial team to set up a meeting and move the conversation forward. Mr. Bruce agreed to sell the right of way across his property in exchange for a one-time payment and a manhole on the property so that he could tie into the line in the future, thus securing the right of way and clearing the major hurdle to the physical sewer extension to the park.

Enigma's negotiations, on the other hand, were not going as well as hoped. They returned to the City of Eureka to discuss annexation into the City for utility service. The City had been made aware of the alternative option with the Town of Ganntt and were not happy about it but also recognized that it needed to protect its service territory from potential outside providers. The City first attempted to block the service by the Town of Ganntt based on its service territory rights with the Regional Council of Governments (RCOG), which had jurisdiction over water and sewer rights in the region. That basically ended in a draw, with the RCOG not truly making a call one way or the other, leaving the situation in limbo. Enigma was starting to get low on time. They had already extended the contract once with Mr. McDoogle and they were within 45 days of the expiration of the extended due diligence period and knew that they wouldn't be able to get another extension. They needed to bring this to a resolution. Enigma discussed the sewer situation with BLA to get an opinion on what their options were. Based on the RCOG decision, there was not a clear finding of jurisdictional rights. The Town of Ganntt suggested annexing into Ganntt and, because the Town had the right to serve any building its in municipality, they believed that it would supersede any existing rights while the property was in the unincorporated county. BLA reviewed the state law on the matter and the RCOG regulations and agreed that the Town of Gannt couldn't be denied service rights if it was in their municipal jurisdiction. Enigma went back to the City of Eureka to see if they could come to an agreement on annexation per Hall County's request.

Enigma met with the City of Eureka for one last time to discuss the situation. Enigma and Best Industrial again reiterated their need for three things: 1) short-term relief from the City's real estate taxes; 2) a waiver of business license fees or a clearly defined ordinance for online sales; 3) the right to bring in an alternative power provider in the event that the power requirements became too large for CPW to serve. Enigma went so far as to state the terms that had been offered to them by the Town of Duncan and the more pro-business environment they had experienced there and their concerns with annexing into the City. Eureka came back to the table and agreed to the following: they would rebate 66 percent of the real estate taxes which would run concurrently with the FILOT agreement provided by the County; they agreed to a sliding scale for business license fees up to 50 percent credit which would offer a credit equal to the pro ration of the jobs above the County average that the company would generate. This wasn't exactly what Enigma had in mind. They could live with the tax reduction but were still concerned about it as the City's millage was the highest in the area, so effective cost was still much higher than being in the Town of Ganntt or Hall County. The business license fee terms were the bigger concern. Enigma felt that it created more questions for prospective tenants who now had to consider their wages when looking into their business license fees and they still felt like fulfilled orders could be considered sales for purposes of the business license.

Enigma determined that their best course of action, after evaluating all the risks involved, was to annex into the Town of Ganntt. While these negotiations were ongoing, BLA had been in discussions with Hall County about the status of the negotiations, their concerns, and the potential of annexing into the Town of Ganntt with an SSRC from the County, as they had previously discussed. The County didn't want to interfere with negotiations between two municipalities within their county, so they stayed neutral during the process. Once Enigma couldn't agree to terms with the City of Eureka, and indicated their intent to annex into the Town of Ganntt, the County re-engaged on potential incentives for the project. From this standpoint, the County and Enigma were aligned as the potential development would be one of the largest ever built in the submarket and could justify the investment for future development. It would also be a substantial investment in the County, which would generate jobs and tax revenue. The County agreed to provide an SSRC for 50 percent of the cost of the sewer and road infrastructure to be rebated over seven years. This allowed Enigma to offset their rate increase with a lower real estate tax cost, allowing for them to maintain the same gross rate when compared with the market.

Closing the Deal

Once the SSRC with the County was agreed to, Enigma had all the necessary agreements and entitlements in place to move forward with the development. Once they had gone through the second reading with the County for the inducement agreement, Enigma entered into an annexation agreement with the Town of Ganntt and subsequently closed on the assemblage. Enigma developed their initial 550,000-square-foot spec building and quickly pre-leased it to a fortune 500 retailer, which expanded the building to over 1 million square feet.

To date, Enigma has developed over 2.8 million square feet of distribution space, which has yielded over 300 new jobs to the area. Since the main sewer line was extended to McDoogle Park, the Town of Ganntt has extended multiple additional sewer branches to other speculative developments which came into the submarket to capitalize on Enigma's initial success. Over 2.3 million square feet of additional development has been built with an additional investment of over $240,000,000 and hundreds of jobs.

All of this was possible due to the initial public–private partnership between Enigma, the Town of Ganntt, and Hall County, who collectively worked together and used public incentive programs that were in place to create a multiplier effect that is still having resounding effects on the community today. The Highway 221 submarket has rapidly become one of the most attractive submarkets in the overall market, with land values and rents skyrocketing. The initial rebates from the Town of Ganntt garnered a significant increase in the tax base between McDoogle Park and the subsequent development, resulting in massive gains to the Town's long-term tax revenue, which the Town has reinvested into additional infrastructure and community development.

Conclusion

After everything is said and done with regards to Engima's project, the key takeaway the reader should understand is that the project was mutually beneficial for all those involved. Even the City of Eureka, which was not selected for annexation, benefited from the development because this was the first project that justified a huge wave of development, both industrial and commercial, in the submarket, much of which would be subsequently incorporated into the City. Due to the size, scale, and immediate success of the project, it justified the submarket for other developers in the market. The road upgrades and utility upgrades were very much needed in the area and the tax base will exist for years after Enigma has exited the market.

This shows the importance of finding good partners and working closely with the local community when developing industrial real estate. It also demonstrates how communities can use their willingness to incentivize up front development to increase their own long-term tax revenue, and how savvy developers can increase the viability of their own projects by partnering with the community versus trying to keep them at arm's length.

Notes

1 The following is a case study of a real industrial development which was developed in the Southeastern United States from 2017 to 2019. Due to the confidential nature of the project and the events herein, the names of the municipalities, law firms, and developers involved have been changed in order to share the story.

2 Jurisdictional creeks and streams, which have to be developed around and cannot be impacted without a nationwide stream permit or purchasing remediation credits.

9 Partnering with Public Agencies to Revitalize Blighted Areas

Joseph Bonora

Introduction

As chapters in this book have shown for many developers, especially smaller-scale developers, public–private partnerships (P3) represent an avenue that can help turn a project from an idea into reality. Most often, a P3 structure can help fill a financing "gap" or address the "but for" component of a potential project, meaning that without the receipt of public dollars or other incentives, the project would not be economically viable.

Public–private partnerships can take many forms, and there are myriad incentives that can be utilized, including tax increment financing (TIF), ad valorem tax abatements or PILOTs (payment in lieu of taxes), impact fee waivers, and forgivable loans to name just a few. All these incentives in one form or the other result directly or indirectly in money in the pockets of the developer, which for a small developer could be what makes a project pencil out and what doesn't. With all this money seemingly out there available for projects, the question becomes: How does a developer get their hands on these incentives?

Obtaining incentives goes well beyond a pretty development plan and an application package; these are just a couple slices of the pie when it comes to getting money from and partnering with the public sector. What is often overlooked is the human factor in the process. For any developer, regardless of the size of their organization, it is important to build rapport with elected officials and public agency staff members and identify which agencies and/or departments oversee economic development programs. Many areas not only have officials who are charged with partnering with developers, but often there are quasi-public–private organizations whose primary role is to leverage their assets to catalyze development and revitalize a neighborhood. One such organization, and one that is gaining notice in the state of Florida, is Community Redevelopment Agencies, or CRAs for short.[1]

What is a CRA you may be asking yourself? It is not the banking regulation imposed on banks to invest locally into the communities where they do business. Rather, it is a physical development structure that invests into a particular designated area. For instance, in Florida, local governments have the authority to designate certain areas, often heavily blighted areas in need of investment, as Redevelopment Areas. Under Chapter 163 of Florida Statutes, when certain conditions exist,

DOI: 10.1201/9781003222934-11

such as blight or a lack of infrastructure, counties or municipalities can establish a redevelopment plan, which includes the overall redevelopment goals for the area and the types of projects to be developed within this area. To carry out the redevelopment plans, CRAs are created and staffed by local governments to administer the activities and programs offered within the redevelopment areas.

This chapter will highlight the benefits of working with CRAs and outline the key steps developers need to take in order to successfully partner with these agencies. Also included in this chapter are three examples of projects we have successfully done in partnership with a CRA in Southwest Florida, which should provide some additional context and hopefully help connect the dots between plan and execution.

Partnering with CRAs

A key fact to remember when working with CRAs, or any other public entity for that matter, is that they want what's in the best interest of the community. They are working with taxpayers' dollars and acting on behalf of the citizens, so any project looking for incentives or public assistance must yield public benefits. As such, prior to approaching a CRA or any other agency with a request for incentives, it is critical for a developer to invest the time necessary to get a full understanding of the goals and needs of the community. All too often, a developer will attempt to *tell* the government and citizens what they want, rather than *ask* them, and then act surprised or appalled when their request for funding, rezoning, etc. is denied.

As was the case with school, even though we all hated doing it, to be successful it really helps to do your homework. In the world of development, this means, among other things, learning the local zoning laws, reading the city's or county's comprehensive plan, reviewing the CRA's redevelopment plan, attending city council or county commission meetings, and meeting with key stakeholders (e.g., community leaders, directors of agencies, neighborhood associations, etc.). Developers will also want to look into the CRA's financials and see what kind of tax increment and other revenue the agency is receiving annually. There is no use in asking for money that the CRA doesn't have and won't be getting. While somewhat labor intensive, this type of diligence will pay dividends and allow you to craft a development plan that will be well received by the community, the CRA, and elected officials, and increase the likelihood of incentives being approved for your project.

In addition to learning as much as possible about local ordinances and development plans, a developer should spend an equal amount of time getting to know the people charged with running the CRA and local government. While the technical side of the business (e.g., zoning, construction, financing) is obviously important, real estate is a relationship business ladened with emotions and motivations that vary from person to person. For the time being, and the immediate future, decisions are being made by people (AI isn't making these decisions just yet), and as any economist will tell you, people are motivated by a drive for incentives. Therefore, understanding what incentives are at play and how they will motivate the decision-makers is a key part of the process.

Once the aforementioned preliminary homework has been completed and the project has reached the stage where a market study has been completed and projections have been prepared, it's time to begin the conversation with the CRA regarding potential incentives necessary to make the project viable. This is where the "but for" test, referenced earlier in this chapter, comes into consideration. Most CRA incentives are designed to fill any financing gap that may exist, so the CRA will be looking for the developer to quantify the shortfall they need the CRA to fund in order to make the project financially feasible. To do this, the developer must prepare a financial model that shows what the project economics (i.e., internal rate of return (IRR), yield on cost, cash-on-cash, etc.) look like with and without TIF and other incentives, and explain why the incentives are needed for the project to yield "market rate" returns.

By way of example, suppose a project would generate a yield on cost (YOC)[2] of 5 percent without TIF based on the developer's projections. Assuming the minimum YOC required by institutional investors is 6 percent, the developer would need the project costs to be reduced or the cash flow increased in order to achieve a "market" YOC necessary to attract equity capital. In this case, the CRA could either provide up-front cash, which could be used to subsidize project costs, or give the developer a rebate of the ad valorem taxes annually, which would increase the property's cash flow. In requesting either of these incentives, the developer will need to show the actual dollar amount necessary to achieve the YOC required by investors, and provide information to support the 6 percent YOC requirement, such as a market survey or appraisal. The more supporting information a developer can provide, the easier it will be for the CRA to get the incentives approved by the CRA Commissioners. So, in this case the "but for" component is that 1 percent between the 5 percent that the project is naturally producing and the 6 percent that is needed by investors, thus without that 1 percent from the CRA the project could not move forward.

While satisfying the "but for" test is a major part of the approval process, it is not the only requirement. Public funds need to yield public benefits, so developers will need to highlight what the community will get out of the deal. Examples of public benefits may include but are not limited to: job creation; sales tax generation and local spending; tourist tax generation; impact fee and ad valorem tax revenue generation; and creation of affordable housing units. For projects we've done in partnership with CRAs, we've hired economists to prepare econometric studies, which estimate direct and indirect job creation, and project the amount of taxes generated as a result of increased local spending, increased property values, and additional tourism. Again, providing this backup information will help the CRA make the case for supporting the developer's request for incentives.

For developers that are successful in checking all the boxes and getting their incentive request approved, the final step is the most important: getting the project built! I stress this point because, all too often, CRAs will approve incentives for projects that ultimately never come out of the ground. In a business where reputation is so important, failing to perform can have serious consequences, and may lead to future requests being denied, or worse. As such, developers should do everything possible to ensure deal execution once all approvals have been received, including lining up their financing and obtaining building permits and other government approvals.

Case Studies

Developers are always looking for a competitive advantage and/or ways to add value and mitigate risk. By utilizing TIF and incentives offered by CRAs and other government agencies, we've been able to pursue opportunities that other developers wouldn't consider and generate above average returns while simultaneously reducing risk. The following case studies should help illustrate how CRAs can assist in getting projects done.

Grand Central: My Introduction to CRAs

In 2016, I was contacted by a broker regarding a vacant 18-acre site in Fort Myers, FL. The site was previously a mobile home park that was cleared and planned to be developed as a condominium project, but the real estate market collapse of 2008 kept the project from coming out of ground. The site had been sitting vacant for nearly ten years and was marketed by multiple brokerages for at least four of those years, but none was able to procure a buyer. Most prospective buyers were retail developers looking to build a big box center, given the site's location on a major thoroughfare, and at one point Walmart was interested in acquiring the site, but none ended up moving forward.

At first, I was skeptical about the prospects for the property. It had been on the market for years but failed to find a taker, so I believed there may be some issues with it, such as environmental contamination or zoning that would limit any redevelopment plans. The property did have great visibility, was centrally located, and was surrounded by both residential and commercial uses, so I decided it was worth a closer look.

After determining that the site was clean and the existing zoning would allow for a number of uses, including multifamily and retail, we entered into a contract with the seller and began working on a development plan. Having spent a good amount of my career working with government agencies at the state and federal level, such as the SBA, USDA, and CDFI Fund, I was familiar with government incentives and was curious if the site qualified for any tax credits or other economic development programs. A quick search on PolicyMap,[3] a website we use to research site eligibility for government funding, demographic information, and other important data points, revealed that the site qualified for New Markets Tax Credits (NMTC), which would allow us to obtain tax credits if we did a building that would result in a business creating employment opportunities. That being the case, I thought there might be some local incentives available for redevelopment, as NMTC areas tend to have higher unemployment rates and poverty levels.

I perused the City of Fort Myers' website and found a link to the Community Redevelopment Agency's site, which included a map of designated redevelopment areas. Seeing that the site I was interested in was located within one of these areas, I called the CRA's office (yes, I called on the telephone; remember this is a relationship business) and spoke with the Director of the CRA. That phone call changed the course of my development career.

During my conversation with the Director, I discovered that the CRA's primary objective was to promote redevelopment, and their main tool for achieving this was an economic development incentive known as Tax Increment Financing, or TIF. She explained that not many developers had utilized the CRA's incentive programs, and she was excited to find out that I was familiar with TIF, tax credits, and other types of government programs. While there was no formal application process, she said she would share the application submitted by the last developer so I could use that as a starting point for my request. This was music to my ears and more than I expected.

The application the Director shared with me was very basic and didn't include much of the information municipalities typically like to see when considering incentives and P3 projects (e.g., economic benefits to the area, job creation, etc.), so I decided to add some of this information to our application to highlight the reasons why the project should be approved and make the Director's job easier.

The effort paid off and the project was approved for a ten-year TIF rebate totaling approximately $4.5 million, as well as stormwater credits, which allowed us to develop the entire site without dry detention or retention areas for stormwater, saving us approximately two acres of land. We broke ground on the project in April 2018 and completed construction in early 2020. The finished product was a 280-unit Class A apartment community, a 3,500-square-foot Krispy Kreme (the first Krispy Kreme in Southwest Florida), an express tunnel car wash, and a 4,600-square-foot multi-tenant retail building.

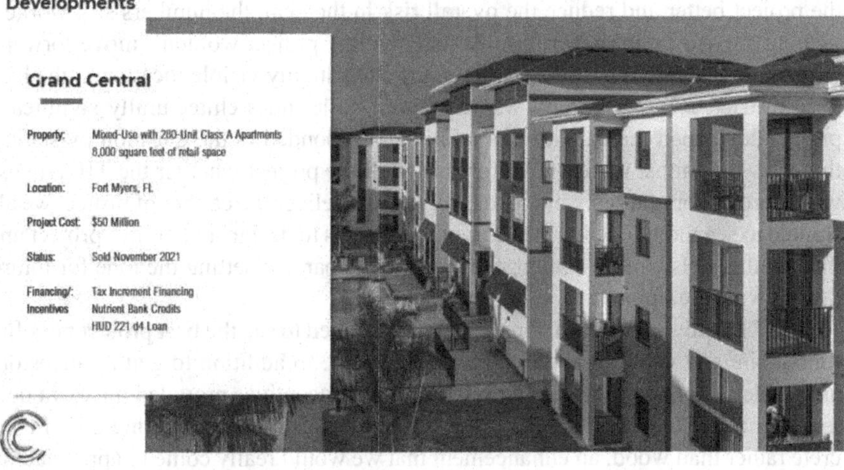

Developments
Grand Central
Property: Mixed-Use with 280-Unit Class A Apartments
8,000 square feet of retail space
Location: Fort Myers, FL
Project Cost: $50 Million
Status: Sold November 2021
Financing/Incentives: Tax Increment Financing
Nutrient Bank Credits
HUD 221 d4 Loan

Figure 9.1 Grand Central.

City Walk

City Walk is a mixed-use project located in Downtown Fort Myers, FL, comprising 376 residential rental units, 13,500 square feet of office space, 9,000 square feet of retail space, and five luxury townhomes. The site was originally approved for the development of a mixed-use property formerly known as First Street Village, comprising 348 condo units and 139,922+ square feet of commercial space, but the project stalled due to the economic downturn and the land was eventually foreclosed on by the bank.

At the time, there were virtually no other apartment communities in Downtown Fort Myers, so we didn't have any rent comps to use for our underwriting. This made it more challenging to raise capital and obtain debt financing. Additionally, the site was less than seven acres, so in order to get the density needed to make the deal viable we would have to build a parking garage. This made the cost to build substantially higher than a typical garden-style project.

In order to help mitigate the higher project costs and lack of rent comps, we worked with the CRA to structure an economic development incentive plan that included a ten-year tax increment rebate totaling $5.5 million and a landscape enhancement grant for $150,000. Once again, I worked closely with the CRA's staff to prepare an application that would highlight the community benefits and economic impacts the project would generate. When I presented the request to the CRA Advisory Board, I was confronted with a number of questions regarding the "need" for incentives by the advisory board members, most of which I was able to answer easily by referencing the projections I prepared and the market data I provided. One question, however, was not as easy to answer: "If you don't receive these incentives, will you still build this project?"

The question caught me a bit off guard because the answer was subjective, and not based solely on data. On the spot, I had to quickly think about the best way to respond to this question. While the incentives would make the economics of the project better and reduce the overall risk in the deal, the numbers still worked without TIF, so I couldn't truthfully state that the project wouldn't move forward without the incentives. However, given the site's highly visible location in the historic district, I knew the city wanted a well-built and architecturally significant project developed there. With this in mind, I responded to the question by stating that we would most likely move forward with the project whether the TIF request was approved or not, but we wouldn't be able to deliver the caliber of project we all wanted to see there. The incentives would allow us to design and build a project unlike anything else in the market, thus raising the bar and setting the tone for future downtown developments.

The CRA advisory board agreed that they wanted to see the best project possible on our site and voted to approve our TIF request. In addition to getting investors and lenders more comfortable with the deal, the incentives provided the extra dollars we needed to design a beautiful project and construct the building out of concrete rather than wood, an enhancement that we would really come to appreciate in 2022 when Hurricane Ian hit Fort Myers.

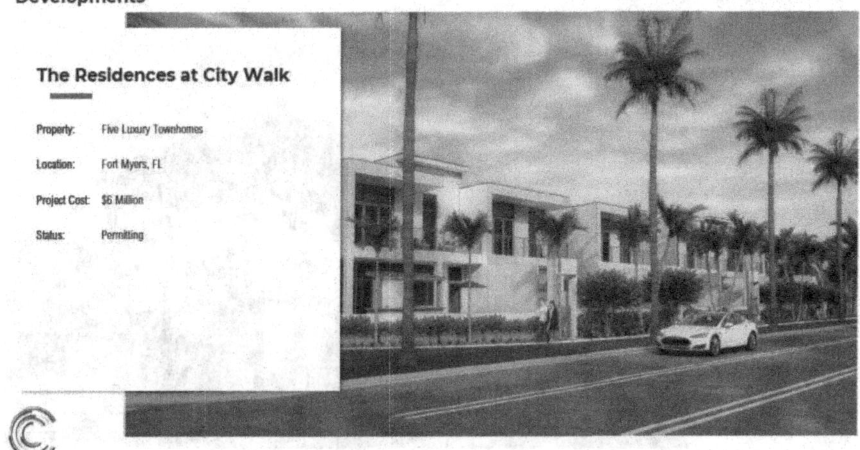

Figure 9.2 The residences at City Walk.

Montage at Midtown

Montage at Midtown is a 321-unit Class A residential apartment community situated on a 4.51-acre site located in the Downtown-Midtown submarket of Fort Myers. The Project is a Class-A property with 15,000 square feet of amenity space, structured parking, condo-quality finishes, and concrete block and plank construction.

The Midtown area in Fort Myers is a 240-acre district just south of the historic Downtown core. The area contains many under-utilized and vacant properties, more than half of which are owned by the City of Fort Myers. Consequently, the Midtown area has become the primary area of focus for the CRA and the city.

Again, given the lack of new development and rent comparables in Midtown, we were faced with the same issues we encountered at Grand Central and City Walk with respect to financing. In addition to these challenges, we were faced with a new wrinkle as it relates to the CRA. In the six years following the approval of the Grand Central TIF request, the multifamily market in Southwest Florida improved significantly, with many projects being developed outside CRA areas without the need for incentives. So why did we still need incentives for Montage?

To help substantiate the need for TIF for our project, we looked at and analyzed the other multifamily projects being developed in Lee County to find out how they were able to pencil out without TIF or other incentives. We determined that the main difference between our project and these other projects was the additional cost associated with high-density urban-infill projects. More specifically, projects being developed on smaller sites in the downtown CBD (central business district) needed parking garages and stormwater vaults, which added approx. $8 million of additional cost.

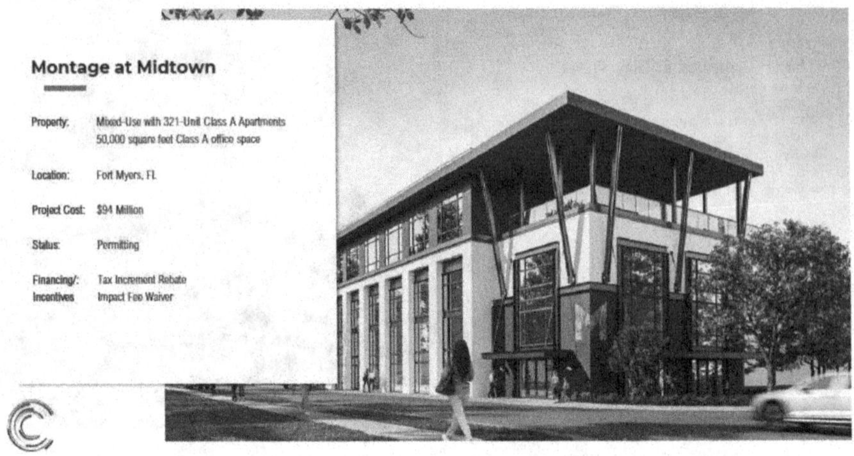

Figure 9.3 Montage at Midtown.

We used this information to support our case for TIF by showing the cost difference between garden and wrap projects and requested an amount with a net present value (NPV) equal to that cost difference. In the case of Montage, that ended up being a 19-year tax increment rebate totaling $11.6 million, which at 4.5 percent equals an NPV of $7.9 million, roughly the cost of the parking garage and stormwater vault. As an added benefit, due to the property's location within an Enterprise Zone, the project qualified for an impact fee waiver, which saved us an additional $2 million.

These incentives have mitigated the impact of rising interest rates (which at the time of this writing have gone up nearly 400 basis points (bps) since our TIF application was submitted) and helped us arrange financing at a time when construction loans are increasingly hard to obtain.

Conclusion

The driving force of this chapter, and what I have tried to get across, is that above all else real estate is a relationship-driven business. The more you can work with people in a personal manner so you are not just a number, the stronger your connection will be, and in turn the chance of receiving more favorable incentives increases. Remember, there are people making these decisions.

In relation to building a relationship, be sure to not over promise and over demand. You must do your homework to understand what is important to the CRA and the community they represent and also what the CRA can actually do for you. But, just as important, you need to inform the CRA what you can do for them; this should truly be a public–private partnership with the emphasis on partnership.

Furthermore, it is always better to under promise and over perform than promise too much and not deliver. So, take the time and work with the CRA to put forth a plan and development that is doable, resulting in a success for both parties.

As the three brief case studies show, the work with the CRA in all three accounts was the deciding factor in getting the development. Without their help, or the "but for," these developments most likely do not happen. Yet it was dependent on myself taking action by picking up the phone, learning about the CRAs, and talking with them. Thus, a key takeaway from this chapter should be, do not be afraid to reach out to your local government official/CRA—they are there to help you. But only reach out after you have done your homework, and when you have something that will help them and the community they represent.

Notes

1 This should not be confused with Community Reinvestment Act legislation, also called CRA, which mandates banks to invest in their communities, although there is some geographic and mission overlap between the two.

2 Yield on cost, also known as the development yield, is the benchmark that investors utilize to assess a project based on its costs and potential return. It is determined by dividing the NOI by the project cost.

3 https://www.policymap.com/

10 Leveraging P3 to Increase the Financial Upside of the Navy's Underutilized Real Estate

A Case Study at Naval Air Station Oceana, VA

Michael Yeary and Stephen T. Buckman

Introduction

The United States Navy currently requires better means and methods to identify underutilized real estate and recommended policy for when and how to attract public–private partnerships to best monetize properties to fill ongoing budget shortfalls estimated for the foreseeable future. These properties have the potential to increase the bottom line for each base's budget. They also present an excellent way for the Navy and the local community have greater connectivity.

There are many factors that encourage the Navy to look towards partnering with local municipalities to development underutilized real estate. While there are a multitude of factors, three primary ones drive the significance of this approach, described in detail below. The first is that Navy budgets are in decline; second, is that Navy budget requirements are growing faster than anticipated due to rapid increases in adversary naval strength; and third, arguably the most important factor is that public–private partnership (P3) developments on underutilized land are a novel approach that allows the Navy to utilize private equity to provide facilities and raise facility revenues while providing developers unique opportunities to target defense clients, contracts, discounted land, and provide for innovative developments that enhance local communities.

Navy Budgets

Navy budgets are in relative decline, and innovation is needed in fiscally utilizing real estate (EUL Fact Sheet). The Navy has limited annual funding to 48–89 percent of estimated facilities sustainment and maintenance costs since automatic budget cutting measures from "sequestration" went into effect in March of 2013 (defense.gov) and risk is taken in facilities to fund higher priorities. This key baseline represents all sustainment funds set aside to maintain existing facilities at their current levels as well as a portion for constructing new building footprints, demolishing inefficient older footprints, and addressing unaccounted for facility costs known as sustainment, modernization, and restoration (SRM) funds. They are separate from appropriated Military Construction (MILCON) funds. Any gap below 100 percent indicates an overall worsening of the state of facilities and their ability to support bases.

DOI: 10.1201/9781003222934-12

Unaccounted facility costs are not a minor factor. For example, the recent Red Hills Fuel Tanks Spill in Hawaii, discovered in 2021, will cost an estimated $2 billion in remediation, subtracting even further from these limited funds, and indicating that a lack of contingency in these funds only exasperates the funding issue (Cocke, 2014). There is no room in the budget for large unplanned expenditures and they come at a cost of already strained funds. The budget is short and the Navy makes it clear in their policy documents they are taking risk on infrastructure spending in order to support higher priority initiatives. Underutilized real estate not only creates a maintenance cost for the Navy, but also an opportunity cost when budgets are much needed elsewhere.

Navy Monetary Requirements

In addition to shrinking budgets, Navy monetary requirements are growing and quickly exasperating budget issues for the Navy. Shortfalls in facilities spending will likely continue as mission-tied funding requirements an increase. On October 6, 2021, Secretary of Defense Mark Esper called for an increased Navy fleet from current day 430 fleet size to over 500 ships within two decades to counter rising state actor threats from China and Russia (Harper, 2020). Chairman of the Joint Chiefs of Staff (CJCS) Milley, an Army General by background, predicted a "bloodletting" in the Army, USMC, and Air Force to fund increasingly important Navy ship expansion in a rare move to reduce one service budget in support of another (McLeary, 2020). General Milley explained that "strategic shifts" require greatly increased spending on Navy priorities to combat modern-day state opponents increasing influence in the Pacific (McLeary, 2020). The recent war between Russia and Ukraine highlights a greater need for Naval power in the Baltics where most Navy bases north of Spain have been decommissioned since the Cold War. World news is providing strong indicators that the need for Naval spending will continue to rapidly increase.

Furthermore, CJCS Milley's vice chairman, Adm. Christopher Grady was recently nominated from the Navy vice other services, a qualitative political shift, indicating that the areas needing greatest focus in today's environment are in the Navy (Baldor, 2021). Facilities directly affect training availability, equipment budgets, ship budgets, and general force readiness within this greater context of effort are becoming rapidly more important. The ability to provide for installations fiscally and potentially leveraging private funding can have a significant impact on the Navy's mission accomplishment.

The Navy and P3

Lastly, Navy installations operate in key locations within our communities and cities. These installations do not just support our active duty service members. They impact communities near bases by greatly affecting regions socially and economically, and steer future development opportunities with both positive and negative externalities. For example, Camp Lejeune contains 32,000 active duty military and has a daily population of 108,000 employed civilian service and contractors (lejeune.mil), which greatly outsizes its surrounding city of Jacksonville, NC that has a population of 72,000. By sheer size, the base is a main driver for the city and, in fact, many cities with large military bases count those bases as primary sectors.

Bases act as local and regional hearts across many levels: economic, social, and community wide. The significant impact of P3 development on perimeter locations of a Navy Installation can address community concerns both on and off base as well as guide the future growth of both. Private innovation is better positioned to understand, market, execute, and maintain semi-private perimeter properties not for military use. Installations typically fence off the outside community as much as possible due to anti-terrorism force protection (ATFP) security requirements, but out-lease opportunities provide P3 partnerships the opportunity to potentially generate mutually beneficial developments for savvy developers, innovative installation support agreements, and have a significant positive impact on surrounding communities.

What Can Precedents Tell Us about Current Military Public–Private Partnerships?

The military has entered into a variety of public–private partnerships over the last 15 years with a strong focus on: utility projects, military housing projects, construction of contractor supporting facilities near and on military bases, and limited use of enhanced use lease authority to allow non-mission-related projects into non-secure and limited secure locations. Further, BRAC strategies, discussed in detail later, to integrate excess land into the surrounding community, and projects that display use of a regulatory plan to guide piecemeal development have been utilized in various capacities. The following precedents outline the size, scope, and use of these typologies and authorities for execution on a military base. These projects provide a proof of concept and lessons learned on how public–private projects can better be executed on military bases.

Utility Agreement Precedents

Davis-Monthan Solar Farm is part of a large number of solar panel projects executed post-2008 as various installations created public–private power purchase agreements to include long-lease structures and guaranteed rates for private companies willing to invest in building, operating, and in some cases providing operations and maintenance to solar farms on military bases. Executive Order 13423 required bases produce their own energy while the 2008 and subsequent government investments provided set aside funding for these projects (E.O. 13423, 2007). Although funding is more limited, there are still energy funds dedicated towards these projects every year and energy use requirements are more stringent in today's environment. Utility contracts have not been limited to just electrical, but also include water and water/wastewater treatment agreements and subsequently privately installed infrastructure both stateside and overseas (e.g., NAS Rota, Spain and NAS Sigonella, Italy). This model provides a potential funding mechanism and legal authority for use on military locations.

Enhanced Use Lease Precedents

10-U.S.C.-2667 (leases: non-excess property) allows acceptance of "in-kind consideration" for any property under the real estate authority of the Navy. This provision is recent and was written into law in 2013 (US Encyclopedia of Law, 2014). Currently, the Navy owns more than 320 million square feet in facilities and 2.2 million acres of land (EUL Fact Sheet: 1) and has a rough estimation of 60 million square feet in "underutilized assets." The Navy has not utilized this authority as aggressively as other services on execution of projects with only two uses at a "large-scale" level as of 2021. There is still great potential across the Navy's portfolio for future projects.

Housing Typology Precedents

Lincoln Housing in North Carolina entered into a 50-year public–private partnership with Camp Lejeune to renovate, maintain, and lease 4,100+ single family and townhome units at six military installations on the coast of North Carolina. Of interest, the Department of Defense (DoD) not only allows Lincoln the use of the land, but it also transferred the facilities to Lincoln for operation and maintenance throughout the length of the lease. To protect Lincoln's interests, waterfall clauses were included to automatically lift restrictive lease covenants if occupancy rates dropped below certain thresholds. For example, in 2015, year-on-year tenancy gaps triggered waterfall clauses and lifted the restriction to rent only to active duty, allowing civil service and retirees to rent on base.

A second lesson learned is the combination of government-provided civic institutes for the success of a project. One key design principle is that housing is located within walking distance of a DoD-operated K-12 school complex, including safe walking infrastructure. Other base amenities of interest are top notch safety practices (base guarded by marines and military police), operation and maintenance (O&M) of utilities and roads, and welfare activities such as a golf course and a beach adjoining the housing areas. Last, Act of God (force majeure) contract clauses are included in the agreement to share risk with Lincoln in case of hurricane, military necessity to reclaim the property, etc., which come at a large price for public sector if executed.

How Is "Underutilized Real Estate" Defined?

"Underutilized real estate" is defined within the Public Contracts and Property Management section of the Federal Code of Regulations. It is defined as:

> an entire property or portion thereof, with or without improvements, which is used only at irregular periods or intermittently by the accountable

> landholding agency for current program purposes of that agency, or which is used for current program purposes that can be satisfied with only a portion of the property.
>
> (41 CFR 102–75.1160)

Key to this definition is that property is not evaluated based on highest and best use, but rather on whether it is executing the "purposes of that agency" at all times vice "intermittently".

This is a broad definition, which allows for a purpose that is fiscal or tied to a specific agency function. It provides one key method of evaluation for determining whether a property is used intermittently for a mission versus full-time use and whether it is a property that is essential to satisfying a part-time mission or if other properties could be used for that temporary function (for example, an annual one-week training exercise).

What Are Available Outgrant Real Estate Instruments and Appropriate Applicable Project Requirements?

Four real estate instruments are available for DoD outgrants and are selected based on the facility type and occupying tenant.

1. *Leases*. These allow a "nonfederal entity exclusive possession of real estate property for a specified term in return for rent or other consideration" (GAO 15-346: 10). For example, an installation may grant a lease for a fast-food establishment on a military base.
2. *Enhanced Use Leases (EULs)*. Similar to leases, EULs are more complex than a normal lease and allow for in-kind considerations or complex long-term agreements. The Red Stone Gateway project is an example, and these are likely more appropriate to P3 complex partnerships in which both government and private entities would utilize land.
3. *Licenses and permits*. These allow entities to use space on an installation for a specific purpose in return for rent or other considerations. Permits are granted typically in exchange for costs of granting the permit such as utilities, maintenance, and services (GAO 15-346: 11). For example, a local farmer's market may use an installation location once a month via a license or a permit may allow them to use the space in exchange for a usage fee.
4. *Easements* work the same as off base and allow for right-of-way access for utilities, roads, or other arrangements as usually made for easements off of a military installation (GAO 15-346: 11).

What Are the Legal Authorities Involved in Out-Lease of Military Installation Property?

Enhanced Use Leasing authority is discussed previously and is legally defined in 10-U.S.C.-2667, 2013. This is the primary authority utilized for out-leasing and

Table 10.1 Applicability of real estate instrument type and asset type to organization utilizing facilities on military installations

Real estate instrument type	Facility type	Nongovernmental and private organizations	State and local governments	Non–Department of Defense (DOD) federal agencies
Leases	Buildings	✓	✓	
	Structures	✓	✓	
	Linear structures	✓	✓	
	Land	✓	✓	
Enhanced use leases (EUL)	Buildings	✓	✓	
	Structures	✓	✓	
	Linear structures			
	Land	✓	✓	
Licenses	Buildings	✓	✓	
	Structures	✓	✓	
	Linear structures	✓	✓	
	Land	✓	✓	
Permits	Buildings			✓
	Structures			✓
	Linear structures			✓
	Land			✓
Easements	Buildings			
	Structures			
	Linear structures	✓	✓	
	Land	✓	✓	

Note: Recommended Real Estate Instrument Type for Facility Type (GAO 13-346).

land leases. It allows bases to be party to a leasing structure, obtain funds from that lease, and directly augment their installation facility budget (NAVFAC, SRM) with those funds.

Second, real estate public–private partnerships are exempt from many federal acquisition regulations (FAR) since 2013. This is significant as FAR regulations are the main guidelines on how the federal government contracts the private sector for supplies and services including construction. A 2013 RAND study indicated that FAR, and related procurement processes, were slowing and making agreements too costly and uncertain for PuP and P3 agreements. Congress responded by providing exceptions to FAR use to allow the DoD to better partner with the public sector to act in a more agile, fast, and partner-like capacity.

Specifically, in 2013, Sec. 331 of the National Defense Authorization Act (NDAA) authorized additional statutory authority for military installations to enter into agreements with local and state governments for "installation support services" (Public Law 112–239, NDAA, Sub D., Sec 331, January 2013). This authority was refined two years later in Sec. 351 of the 2015 NDAA providing contractual clarification on PuP cooperation (essential if leasing customers are government-contracted service providers) and transferring authority for these agreements to fall under the Real Property section of U.S. Code; in affect making it so military installations are not required to follow Federal Acquisition Regulations (FAR).

What Is the Policy and History for Disposal of Excess Real Estate on Military Installations?

Military installations that identify real estate as excess, i.e., of no military value to the base or its missions, are to transfer that property to the General Service Administration (GSA) for potential disposal through the Base Realignment and Closure process (BRAC). Installation commanders gain no benefit from this process and any funds generated are sent to the treasury via the GSA. Further, reductions are a highly political issue and policy combating "encroachment" is key to ensuring the military's mission is protected from real estate restrictions (NAVFAC BMS). Since the original 1990 Defense Base Realignment and Closure Act, six rounds of BRAC have been performed, affecting 350 installations with 45 full installation closures occurring since 2005.

No BRAC Commission has made a recommendation for additional closures since 2005, and as the country enters into an era of great power competition (Congressional Research Services, 2021: 2; Principi, 2015) it is increasingly unlikely that this process will be utilized in the near term. However, there is a "stealth BRAC" process that allows smaller reductions under the BRAC act via Executive fiat (Principi, 2015). Annually, the DoD submits through the Secretary of Defense for Presidential approval a list of more minor reductions and realignments which can be accomplished via Executive Fiat, i.e., with little congress interaction.

What Navy Installation Requirements Are Key to the Planning Process?

Similar to municipal planning, all Navy installations are required to have a master plan that addresses real property master planning, sustainable design and development, short- and long-term range planning, transportation planning, and environmental planning (10 USC 2864). One layer down, the United Facilities Criteria (UFC) outlines master planning requirement in UFC 2-100-01 and includes provisions for area development planning, network planning, potential use of form-based planning, and various other standard planner tools. Within that UFC, installations are instructed to "conserve land as a resource to be utilized to the maximum extent possible" via compact development (2-100-01(2–2.1)) and utilize infill development (2–2.2), transit-oriented development (2–2.3), consider horizontal and vertical mixed-use development, discourage one-story development, and take into account "healthy community planning practices . . . community gardens. . . [and] walking, running, and biking" (6–27). All of these provide the basic framework for a mixed-use urban design strategy with any development within a base.

Case Study: Naval Air Station Oceana

A case study on how the Navy can improve its RFI (request for interest) process to engage in P3 was performed at Naval Air Station Oceana (NAS Oceana). The site was selected due to publically available information, proximity to an urban area, availability of underutilized perimeter real estate, and typical base mission areas.

NAS Oceana is located in close proximity to Virginia Beach, VA and has a long history as it was acquired in 1940 as part of the World War II buildup. The installation contains standard facilities to man, train, and equip deployable forces and contains one of two primary mission areas for Navy installations, an airport (port, airstrip, or both are typical medium and large size installation primary capabilities). Last, it is located in the dense community of Virginia Beach, providing potential marketability of leased parcels, although the immediate surrounding area is mostly residential with little commercial real estate except on the dense north end.

Virginia Beach City, Virginia is a historic coastal resort town known for tourism, military support, and prime real estate along the Chesapeake Bay. It is the largest city in the state of Virginia with a population of 459,470 (2020 Bureau of Labor Statistics; US Census Bureau, n.d.). Since 1979, there has been a strong movement to preserve green space along southern portions of the city to ensure continued preservation of the Back Bay wetlands area while development occurs in the northern areas of the city including the areas immediately adjacent to NAS Oceana.

Virginia Beach is best known for its boardwalk beach area. This iconic three-mile stretch of beach in the downtown resort area is the longest continuous public beach in the United States and is lined with hotels and restaurants, with a walkable environment. Much of the rest of the area is conventional suburban design, including over 140,000 single family houses in multiple neighborhoods. Virginia Beach's largest employer is Naval Air Station Oceana, which is located in prime real estate along the primary access to the Virginia Beach resort area, I-284. The installation is ideally located 2.4 miles from the resort area and 2.9 miles from the beach.

Project RFI Proposal

NAS Oceana has six underutilized parcels with available public data that were used for this study and potentially much more in unreleased data. NAS Oceana released six separate RFIs to advertise for interest from developers on July 1, 2020 for underutilized land proposals to engage the private side for best ideas on how to utilize parcels along the base's southern perimeter. The RFIs were posted on FedBizOpps.gov and responses were required within 14 days, including the July 4 holiday. No consideration was provided for preparation costs of LOIs (letters of intent). There was no other extensive marketing and this is a non-typical item to be posted on the site. Requirements were overly broad and there was no clear indication of what the Navy envisioned in terms of a P3 partner use of the land. No responses were received. As of Spring 2021, the project is under revision while land continues to go underutilized, maintenance continues to be base funded, and the RFIs are reworked for potential future re-solicitation.

Simple business procedures could improve this process dramatically by giving ample time for response, communicating a clear vision and appropriate restrictions in the RFI, finding an appropriate way to post a non-typical RFI project, and engaging the business and municipality urban regime in advance in a search for P3 partners.

One primary reason for poor responses was a lack of government expertise in how best to market the RFI. Examining the RFIs revealed unclear project purpose, unclear security requirements, unclear encumbrance requirements for air operations (such as building heights, noise, cranes, etc.), and infers that an offeror would need to fund additional environmental impact studies in order to develop on the site with no risk-sharing for the government.

Best practices would be to alleviate these concerns as part of site preparation prior to the RFI. The RFI can continue to be a simple document but needs to clearly communicate restrictions as part of any solicitation from the private community. Further, the specific out-lease instrument envisioned, such as a public–private partnership via EUL, is not mentioned. Offerors had little ability to infer the government was seeking a P3 partnership and EUL in-kind consideration in exchange for partial use of underutilized properties.

In the original RFI the Navy provides 17 potential uses but does not distinguish or provide a clear vision for the Navy desires. Restrictions such as helicopter flying zones and specific environmental areas that may be impacted or provide risk to development are not clearly defined. Developers were likely unclear on desired land use by the government with such a diverse set of options, lack of restrictions, and no communicated project vision.

In order to attract private investment and ideas for the highest and best use of the land, it is important for the government to better understand the market to determine an initial project vision and communicate to private real estate developers their intent for project proposals (pre-negotiation agreement), perform full site preparation, give adequate time for proposal responses, include local municipalities as potential partners, and clearly identify any site encumbrances for development—all of which are expert areas of private real estate partners.

The Navy has not fully outlined policy and process milestones for procuring an EUL. However, the United States Airforce (USAF) has a very developed process, based on a standard solicitation and transition into portfolio management phase. This study revealed the Navy should develop an appropriate process for its similar rules and restrictions, but include phase 0 project identification steps that allow for P3 best practices. Specifically, as shown in Chapter 1 in this book, Friedman's (2016) best practice P3 process includes additional steps of: initial site analysis, market analysis, and an updated RFI that can better garner private or municipal interest for a partnership. The Navy can utilize USAF's existing model and include market analysis, site analysis, and municipal coordination prior to an RFI to identify potential P3 partners. All three should be performed in Phase 0, project identification, to allow for a P3 partner's expertise to assist in execution of Phases I–III and have the greatest effect on developing appropriate programming for a development based on private market expertise.

Based on the previous studies, a draft RFI is provided, adjusting the previous 2020 solicitation and including key relevant information gained from the case study for better developer response (Appendix 1).

Pilot Key Process: RFI Development at NAS Oceana

The RFI is the key engagement piece the Navy needs to engage and propose a vision to potential P3 partners. The RFI is the key formal process to advertise interest in partnering for a project in the initial development phase prior to any formal solicitation and to identify potential partners for further negotiation and interest. No proposal is developed by an interested party at this phase, but an initial vision is provided. Although it is a general interest statement, it is important to let potential partners know the government's intentions and flexibility as well as key parcel data to engage private partners for developing an EUL agreement.

As part of the study, prior to sending the RFI, REIS software, GIS, military base property use, and a Virginia Beach comprehensive plan review were performed on NAS Oceana's perimeter real estate with go/no-go criteria to determine underutilized real estate that could be potentially offered via RFI for P3 development (Figure 10.1). Based on the review, key sites were identified and applied to create the composite of underutilized real estate: wetlands, strategic growth areas, circulation patterns, market analysis opportunities, military use of the property for potential underutilization, and ability to relocate fence site lines if needed to allow for public access. For the purposes of this pilot, six of these parcels whose key information had already been released to the public via prior release were available for academic study, and were examined for potential as part of an out-lease analysis along the perimeter of the base.

Composite of Potential Underutilized Locations

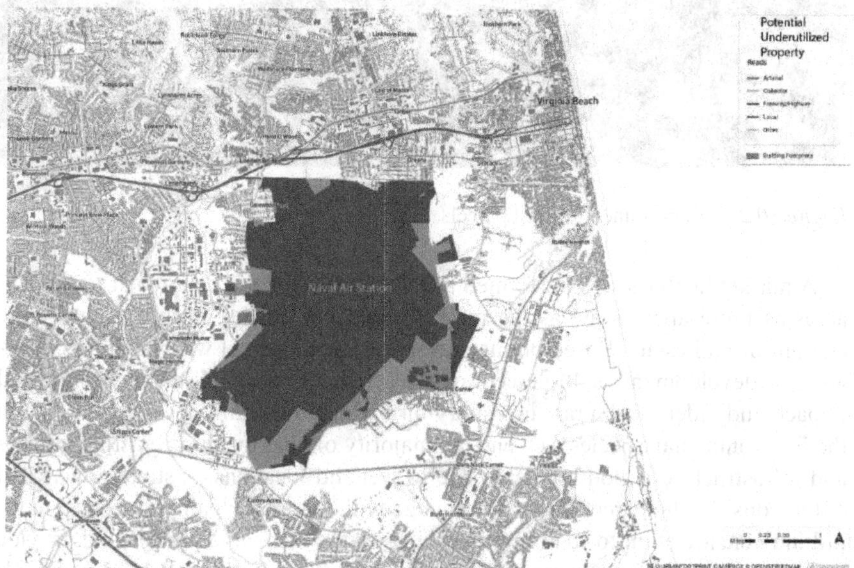

Figure 13, Composite of Restricted Real Estate and Potential Good Targets for Underused Land

Figure 10.1 Restricted real estate.

Table 10.2 Acres in play

#	Parcel	Acres Total	Acres Wetlands	Acres Usable
1	Commissary West	29.9	4.2	25.7
2	Commissary East	24.5	1.1	23.3
3	Harpers Rd / AG Field	19.5	0.6	18.8
4	Harpers Rd / Old Housing	55.2	0.0	55.2
5	Oceana Blvd / AG Field	200.4	110.0	90.4
6	Owl's Creek	187.9	100.0	87.9
	Total	517.3	216.0	301.3

Figure 10.2 Oceana's underutilized parcels.

A market analysis was performed and indicated a phased approach to the 517 acres as more appropriate as the market could not quickly absorb such a large amount of real estate. Based on the site analysis, one parcel was much less desirable for development as 40 percent of the acreage is wetland, the environmental impact study identified a rare migratory bird which may limit development due to the Environmental Species Act, and the majority of land touching existing utilities and infrastructure is non-buildable due to wetland locations. A strategy was laid out to consider the parcel for land exchange with the municipality in exchange for in-kind consideration to fund improvements to the other five parcels and provide necessary fencing improvements to move the base boundary and prepare parcels for out-lease. The market study and site analysis resulted in an initial negotiation position that may allow a win-win with the municipality in executing preservation of green space and seek to generate $2.5 to $3.5 million dollars in fence improvements. The RFI includes phased language and would exclude this parcel if the local municipality were interested in a land exchange for in-kind consideration.

One area specifically left out of the RFI, Appendix B, is the amount of programming envisioned from the market analysis. Although REIS software was used, the government is largely bad at estimating marketability and any included estimates are more to give a contractor a rough idea rather than specific requirements for development. Instead, a project vision section is included in the proposed RFI to give the developer a ballpark idea for the property vision the Navy would find more desirable for P3 partnership. Further, the initial market analysis is included as an enclosure to provide better initial information to generate interest. Most developers will quickly be able to generate more robust products, however the RFI is to initiate initial ideas and partnerships and is not designed to encumber the private sector with expenses prior to engagement.

The RFI combines parcels into one RFI rather than six separately posted documents. This streamlines communication and allows for developers to target portions of parcels on both sides of Oceana Boulevard. It also reduces confusion from a federal RFI posting system. The RFI includes market and site analysis, and recommended base desires for either in-kind consideration are already predetermined from the MEP examination. Focusing on seeking P3 partners via RFI, use of market and site analysis products, and smart selection of underutilized property all are key to risk-sharing for military partners to engage with local developers and municipalities.

Policy Recommendations

The Navy should evaluate its perimeter real estate for stateside locations by contracting site analysis, market analysis, and base interviews to generate an underutilized real estate parcel location list as well as supporting data for soliciting private developers. Maintenance of the list can be conducted through public works biannual review and collected by Navy echelon III real estate departments for inclusion in a larger database of potential out lease opportunities. Echelon II and III NAVFAC units should consider utilizing this list to help inform the potential for military uses via EUL for large programmatic requirements while Echelon IV (installation level) can utilize the list for P3 private development in exchange for in-kind consideration or lease income. The Army reserve currently maintains a similar database in conjunction with the GSA and serves as a model on how to better manage identified underutilized areas.

Third, prior to enactment of any policy change, it is recommended that NAVFAC contract a similar site and market analysis to study perimeter locations across three to five additional installations in order to ensure this approach works at domestic locations. A medium-sized air base was studied for the NAS Oceana pilot. Small and large installations, overseas and domestic, as well as port or R&D-focused installations would make good candidates to test case the validity of utilizing an analysis process as a means and methods of identifying potential underutilized real estate for P3 development. This may change the approach to those installations; for example, overseas installations may not be an appropriate target for EULs due to legal requirements or may just need altered guidance for enactment.

Additionally, all services did a poor job engaging the private sector P3 development of property despite heavy P3 engagement in many other areas such as base installation maintenance. Recent law changes have removed many of the traditional restrictions for the success of these partnerships, including very restrictive FAR and anti-terrorism force protection requirements. In addition to adopting either the USAF or USACE process for EULs, the Navy should also include P3 engagement prior to the RFP (Requests for Proposal) step of government solicitation, as shown in the pilot test case. Success or failure should be monitored and measured by the amount of private engagement seen in EUL projects. Once successful, secretaries of the Navy, Air Force, and Army should consider sharing P3 best practices across all domestic facility policies.

P3 best practices with other public entities have been successful with much earlier engagement and Friedman's recommended practice of site predevelopment. The market analysis and site analysis, as well as preparing environmental permitting requirements, cost estimating perimeter security change requirements, and communicating a clear initial vision with the municipality and potential developer partners all increase the marketability of a potential development. Because the Navy has performed so few of these projects, it will take additional engagement to educate the private sector on how best to find these opportunities. Underutilized real estate databases should be made available through a separate platform for review.

In summary, it is recommended that the Navy develop EUL policy that includes solicitation methods via P3 best practice and directs installations to inventory and manage underutilized real estate by contracting site and market analysis, coordinate military requirements from identified target underutilized real estate with base public works staff, collect and manage regional lists at Echelon III higher level units for use in steering larger programmatic requirements, and submit this policy for implementation by the SECNAV for service-wide execution.

Conclusion

The Navy estimates it owns 60 million square feet of underutilized assets (EUL Fact Sheet) and is currently funding research into how better to approach and enter into EULs and P3 partnerships. The DoD recognizes the potential for using underutilized real estate via enhanced use leases which allow the DoD the opportunity to reduce facility costs by leasing non-excess, underutilized military real property for in-kind consideration. Budget shortfalls continue to put pressure on base Commanders to find creative solutions to funding facilities. Within this framework, private partners are better positioned to bring their expertise in leveraging funding sources traditionally unavailable to DoD facility managers in exchange for gaining access to unique properties in highly valued locations around military installations nationwide.

The pilot at NAS Oceana was developed as a test case for how the Navy can identify underutilized real estate on a base using site analysis and real estate development tools and how to apply Friedman's (2016) principles to develop RFIs that lead to P3 partnerships and later long-term out-lease arrangements beneficial to both parties. Removal of legal restrictions in the 2013 and 2015 NDAA are key to allowing the Navy to become a strong P3 partner. The study also shows where the DoD is currently lacking in best process for engaging P3 partners for EUL work. The sites in question are being repackaged as of Spring of 2021. There is potential for savvy developers and municipalities to partner with local military bases and help them in proposing visions for underutilized property through out-lease opportunities.

Appendix

Appendix A

Economic Incentive Commitment Agreement

ECONOMIC INCENTIVE COMMITMENT AGREEMENT

This Economic Incentive Commitment Agreement (hereinafter "**Agreement**"), is made and entered by and between Cobb County, Georgia (hereinafter "**County**"), a political subdivision of the State of Georgia, Thyssenkrupp Elevator Corporation, a foreign corporation authorized to conduct business in the State of Georgia (hereinafter "**TK Elevator**") and Thyssenkrupp Real Estate North America, LLC a foreign limited liability company authorized to conduct business in the State of Georgia (hereinafter "**TK RENA**") (collectively, TK Elevator and TK RENA are "**Thyssenkrupp**"), and BRED Co., LLC (hereinafter "**BRED**"), a Georgia limited liability company, with a local principal address of 755 Battery Avenue, Atlanta, Georgia 30339,

RECITALS

WHEREAS, the County has enacted the Economic Development Incentive Ordinance of Cobb County to provide for quality, controlled growth, retention, redevelopment, and rehabilitation of targeted areas within the County; and

WHEREAS, Section 2-173 of the Cobb County Code (the "**Code**") sets forth the County Special Economic Impact Incentive Program ("**Program**"), the purpose of which, pursuant to Section 2-173(a), is to provide services and business assistance to businesses in relocating their businesses to the County or expanding their business in the County; and

WHEREAS, Section 2-173(b)(1) of the Code sets forth certain eligibility criteria required of businesses to qualify for the Program, including serving as headquarters or being engaged in manufacturing and/or emerging technologies/industries; and

WHEREAS, Section 2-173(b)(2) of the Code sets forth additional criteria that eligible businesses must meet to qualify for the Program, including, at least two of the following: the addition of at least 150 new jobs to the County, the payment of an average salary of at least 1.25 times the County average for that industry, and/or the investment of $30,000,000.00 or more in the County; and

WHEREAS, Thyssenkrupp, a German-based conglomerate that manufactures and sells elevators, commits to: (a) relocate the regional headquarters and emerging technologies business of Thyssenkrupp Elevators Americas to an approximately $130,000,000 office building ("**Office Building**") to be constructed by BRED; and (b) to build an approximately $101,000,000, 420-foot tall state-of-the art elevator qualification and test tower ("**Elevator Tower**"). The Office Building and Elevator Tower will be located at Circle 75 Parkway, Atlanta, Georgia, Parcel number 17091400110 (the "**Project Site**") (both structures collectively constituting the "**Project**"); and

WHEREAS, Thyssenkrupp commits that the Project will result in the creation of approximately 125 new jobs in year one of operation, with an anticipated increase to 863 new jobs within the 10 year incentive period ("**Incentive Period**") at an average salary of $94,000 at full ramp up for all roles; and

WHEREAS, Thyssenkrupp estimates that the direct and indirect investment in the County as a result of the Project is expected to be approximately $237,931,614; and

WHEREAS, Thyssenkrupp commits to purchase $16,400,000 in new furniture, fixtures, and equipment for the Office Building and the Elevator Tower; and

WHEREAS, BRED and Thyssenkrupp, as developers of the Project Site, have agreed to construct each facility, and as such to benefit from economic incentives in accordance with the agreement that Thyssenkrupp will lease/occupy the specified space in the Office Building, and construct, own and occupy the Elevator Tower upon completion; and

WHEREAS, the County, through the appropriate authority of the Board of Commissioners, and under the Cobb County Code, agrees as provided herein to provide certain incentives available for development of the Project with the understanding that such incentives are extended to assist Thyssenkrupp's relocation to the County; and

WHEREAS, Thyssenkrupp and BRED each have submitted applications to the County seeking consideration of incentives available (the "Applications" attached hereto and referred to as Exhibit "A"); and

WHEREAS, in accordance with procedures adopted by the Board of Commissioners, the Economic Department Division Manager (**"EDD"**) analyzed the Applications utilizing the County's economic impact model, reached out to departments regarding potential incentives, determined that the Project was potentially eligible for incentives, arrived at a staff recommendation, and obtained the review and comments of the Finance Department as to its incentive recommendations; and

WHEREAS, the EDD Manager made a recommendation to the County Manager, District Commissioner and Chairman of the Board of Commissioners and received their concurrence for the incentives offered; and

WHEREAS, the Chairman of the Board of Commissioners issued offers of incentives in accordance with the recommendation of the EDD Manager following concurrence pursuant to those certain offer letters from the Chairman to BRED, dated July 26, 2018, and from the Chairman to Thyssenkrupp, dated July 26, 2018 (collectively, the **"Offer Letters"** and the Offer Letters are incorporated herein by reference and made a part of this Agreement); and

WHEREAS, Thyssenkrupp and BRED have agreed to accept the incentives offered;

NOW, THEREFORE, in consideration of the mutual covenants contained herein, the parties agree as follows:

1. **PROJECT**

A. Thyssenkrupp – Thyssenkrupp agrees to construct a $101,000,000, 420-foot tall Elevator Tower on the Project Site (the **"TK Elevator Tower Requirement"**). Construction of the Elevator Tower is expected to be completed by 2021.

Thyssenkrupp agrees to create 125 jobs in year one of operation and anticipates creating an additional 738 permanent full-time, or full-time equivalent over the Incentive Period (the **"TK Jobs Requirement"**).

Thyssenkrupp agrees to occupy a portion of the Office Building. Thyssenkrupp employees will also occupy the Elevator Tower (collectively, the **"TK Occupancy Requirement"**).

Additionally, Thyssenkrupp agrees to purchase new furniture, fixtures, and equipment to be used in the Office Building and the Elevator Tower for approximately $16,400,000 (**"FF&E Expenditure"**).

B. BRED – BRED agrees to construct an approximately $130,000,000, approximately 290,000 square foot Office Building on the Project Site (**"BRED Office Building Requirement"**), a portion of which will be leased by Thyssenkrupp, either directly or through an affiliate, for use as its corporate headquarters. Construction is expected to commence in the 4th quarter of 2018 and be completed by 2021, with Thyssenkrupp expected to occupy its leased portion of the Office Building after completion in 2021.

2. **INCENTIVE PERIOD**

The parties agree that the Incentive Period shall commence on January 1 of the year following the issuance of the Certificate of Occupancy for the Office Building and Elevator Tower and remain in effect for a ten-year period.

3. **VALUE OF INCENTIVES**

The County agrees to authorize an incentive package valued at approximately $1,031,000 for the Project, based on the following:

i. **Office Building Permit Fee Cap** – Pursuant to Section 2-173(c)(4) of the Code, the building permit fee for the Office Building to become Thyssenkrupp's headquarters shall be capped at $5,000.00. The building permit fee is normally calculated at $6 per $1,000 of construction value, and the parties agree that the building permit fees for the Building would ordinarily cost $600,000. Consequently, the value of this incentive is $595,000.

ii. **Elevator Tower Building Permit Fee Cap** – Pursuant to Section 2-173(c)(4) of the Code, the building permit fee for the Elevator Tower to be constructed, occupied and operated by Thyssenkrupp shall be capped at

$5,000.00. The building permit fee is normally calculated at $6 per $1,000 of construction value, and the parties agree that the building permit fees for the Elevator Tower would ordinarily cost $606,000. Consequently, the value of this incentive is $601,000.

iii. **Annex Building Permit Fees** – Pursuant to Section 2-173(c)(4) of the Code, the building permit fee for interior improvements shall be capped at $5,000 for the Annex. The building permit fee for the Annex would be approximately $9,375. Consequently, the value of this incentive is $4,375,

iv. **Water System Incentive** – Pursuant to Section 2-173(c)(3) of the Code, a water systems incentive package shall consist of a combination of reduction in line fees, meter fees, backflow prevention costs, and/or installation/upgrade of required water/sewer infrastructure in the total amount of $100,000 for the Office Building and $16,000 for the Elevator Tower.

4. **ANNUAL REVIEW; REPORTS**

Upon request by the County, but no more than once (1) annually after completion of construction, Thyssenkrupp and BRED shall jointly submit to the County an annual progress report documenting the number of new jobs at the Project Site and any other information relevant to this Agreement that the County reasonably deems appropriate for the Incentive Period (no later than sixty (60) days after such request). Failure to report, as required by this paragraph, will not be deemed to be a violation for purposes of Paragraph 5 unless BRED and/or Thyssenkrupp have failed to file such report within thirty (30) days of the receipt of a written notice of such failure.

5. **RECAPTURE OF INCENTIVES**

Thyssenkrupp agrees that should it fail to substantially satisfy the TK Elevator Tower Requirement, the TK Jobs Requirement, the TK Occupancy Requirement or the FF&E Requirement described herein during the period for which the incentives are granted, it will immediately reimburse the County for the full value of any and all incentives received hereunder. BRED agrees that should it fail to satisfy the BRED Office Building Requirement, it will immediately reimburse the County for the full value of any and all incentives.

If any incentive provided to either BRED or Thyssenkrupp is later determined to be illegal, unenforceable or invalid, BRED and Thyssenkrupp, individually, agree to promptly reimburse the County for the full value of any and all incentives provided hereunder.

6. **BREACH, NOTICE & TERMINATION**

If either BRED or Thyssenkrupp materially fail to fulfill its obligations under this Agreement, including the covenants in Paragraph 8, the County shall provide written notice and thirty (30) days to cure the breach or to show cause why the party should not be deemed to be in default. If the breach is not corrected within the thirty (30) days after the written notification, and no resolution or mediation of the breach has been reached, the County may terminate or suspend the incentives granted under this Agreement, and may require the payment by BRED or Thyssenkrupp of the amount of incentives that would have been payable had the property not received said incentives. In the event of such termination or suspension, the County may pursue all appropriate legal remedies and/or relief.

7. **NOTICES**

Except as may otherwise be provided herein, all notices, demands, requests and other communications under this Agreement shall be in writing and shall be either personally delivered, sent by registered or certified mail, or sent by courier, to the following addresses (or to such other address as may be designated by written notice transmitted in accordance with this provision):

In case of the County, to:
Economic Development Division Manager
Community Development Agency
P.O. Box 649
Marietta, Georgia 30061
(770) 528-2018

In case of Thyssenkrupp, to:
Thyssenkrupp
Thyssenkrupp Elevator Corporation
c/o thyssenkrupp North America, Inc.
111 West Jackson Blvd, Suite 2400
Chicago, Illinois 60604
Attn: John H. Anderson, Head of RE North America

Or

CBRE Advisory & Transaction Services
Attn.: Jack DiNardo
321 North Clark Street, Suite 3400
Chicago, IL 60654
312-861-7817
Jack.DiNardo@cbre.com

In case of BRED, to:
BRED Co., LLC
Attn: Mike Plant
President and CEO
755 Battery Avenue
Atlanta, Georgia 30339

404 -614 -2191
Mike.plant@braves.com

8. **FALSE STATEMENTS, PENALTIES**

Thyssenkrupp affirmatively covenants that no false statements have been made to the County by Thyssenkrupp or its representatives in the process of obtaining approval for the economic incentives. If any representative of Thyssenkrupp has knowingly made a false statement to the County to obtain the incentives or fails to redress a false statement that was unknowingly made, the County may require Thyssenkrupp to immediately return all benefits received under the Agreement and Thyssenkrupp shall be ineligible for any future economic development assistance from the County.

BRED affirmatively covenants that no false statements have been made to the County by BRED or its representatives in the process of obtaining approval for the economic incentives. If any representative of BRED has knowingly made a false statement to the County to obtain the incentives or fails to redress a false statement that was unknowingly made, the County may require BRED to immediately return all benefits received under the Agreement and shall be ineligible for any future economic development assistance from the County.

Any person who provides a false statement to secure economic development assistance may face criminal charges.

9. **INDEMNIFICATION**

Cobb County shall not be liable for any contractual or other claim caused by or related to the construction, implementation, and/or operation of the Project.

BRED agrees to indemnify and hold harmless Cobb County from any claims or actions arising in any manner related to the construction and/or operation of the Office Building (except for those related to Thyssenkrupp's construction of the interior of the premises or occupancy of the premises that it leases in the Office Building), including specifically any claims for negligence, accident or injury caused by any party, its agents, employees, or contractors, and non-performance and non-payment of any obligation or debt incurred by or on behalf of BRED relative to any aspect of the Office Building.

BRED agrees to indemnify and hold harmless Thyssenkrupp, its agents, affiliates, employees, officers, directors, members, partners and shareholders from and against any and all liabilities, judgments, demands, causes of action, claims, losses, damages, costs and expenses, including reasonable attorneys' fees and costs, arising from any breach or default under this Agreement by BRED.

Thyssenkrupp agrees to indemnify and hold harmless Cobb County from any claims or actions arising in any manner related to the construction and/or operation of the Elevator

Tower (and those related to the interior construction and/or occupancy of the premises that it leases in the Office Building), including specifically any claims for negligence, accident or injury caused by any party, its agents, employees, or contractors, and non-performance and non-payment of any obligation or debt incurred by or on behalf of Thyssenkrupp relative to any aspect of the Elevator Tower.

Thyssenkrupp agrees to indemnify and hold harmless BRED, its agents, affiliates, employees, officers, directors, members, partners and shareholders from and against any and all liabilities, judgments, demands, causes of action, claims, losses, damages, costs and expenses, including reasonable attorneys' fees and costs, arising from any breach or default under this Agreement by Thyssenkrupp.

These indemnifications shall survive termination of this Agreement.

10. **ENTIRE AGREEMENT; NOT BINDING PRIOR TO EXECUTION**

This Agreement (together with the Offer Letters as expressly incorporated herein) contains the entire agreement between the parties with respect to the Program for the Project, and no promise, terms, or obligations, other than herein set forth, or subsequently set forth in writing and signed by all parties hereto, shall be binding upon any party hereto. This Agreement supersedes all prior negotiations, representations, or agreements, whether written or oral with respect to the Program for the Project. This Agreement may be amended only by written instrument, subject to approval by the Cobb County Board of Commissioners and execution by all parties.

11. **ASSIGNMENT**

The parties may not assign this Agreement or the rights and obligations herein without the consent of the other parties. The Agreement shall be binding upon and inure to the benefit of and be enforceable by the parties to the Agreement and their respective successors and permitted assigns. Notwithstanding the foregoing, the parties hereto may freely assign its rights and obligations hereunder to any of its affiliates.

12. **GOVERNING LAW**

The parties agree that economic incentives are governed by the Official Code of Cobb County, Georgia, Sec. 2-166, et seq., as amended and this Agreement. The parties further agree this Agreement shall be governed exclusively by the internal laws of the State of Georgia, without regard to its conflicts of laws rules. The courts located in Cobb County, Georgia shall have exclusive jurisdiction to adjudicate any dispute arising out of or relating to this Agreement. Each party consents to the exclusive jurisdiction of these courts.

SIGNATURES TO FOLLOW ON THE NEXT PAGE

IN WITNESS WHEREOF, the parties acting through their duly authorized agents have caused this Agreement to be signed, sealed and delivered for final execution by the County on the dated indicated herein.

THYSSENKRUPP REAL ESTATE
NORTH AMERICA, LLC

By: ____________________________

Name: __________________________

Title: __________________________

Date: __________________________

Attest: __________________________

By: ____________________________

Title: Notary

BRED CO., LLC

Seal

By: ______________________

Mike Plant, Executive Vice President

Date: September 18, 2018

Attest: ______________________

By: ______________________

Title: Notary

JONATHAN E. SMITH
MY COMMISSION EXPIRES
NOTARY
PUBLIC
DECEMBER 4, 2018
FULTON COUNTY, GEORGIA

Seal

COBB COUNTY, GEORGIA

By: ___________________________
Michael H. Boyce, Chairman
Cobb County Board of Commissioners

Date: ___________________________

Attest:

____
By: Pamela L. Mabry
Title: County Clerk

Approved as to Form:

____
Deborah L. Dance
Cobb County Attorney

Seal

EXHIBIT A

APPLICATIONS

TO BE ATTACHED

BUSINESS INCENTIVE APPLICATION

(Select one)

- ☐ ***Entrepreneurship & Innovation***
- ☐ ***Small Business & New Start***
- ☐ ***Business Retention***
- ☐ ***Targeted Industry***
- ☑ ***Special Economic Impact***

~Instructions~

Please fill in the following sections:

- ***This Page** with the box checked for the incentive for which you are applying*
- ***Applicant Information***
- ***Applicant Certification***
- ***Business Incentive*** requirements for the incentive for which you are applying
- ***Provide required additional required information***

Please return the completed application and attachments *(if applicable)* to:

Cobb County Economic Development Division
P.O. Box 649
Marietta, Georgia 30061-0649

Or scan and email to: econdev@cobbcounty.org

Upon receipt of a completed application, Cobb County may require other information as deemed appropriate for evaluating the financial capacity and other factors of the applicant. Cobb County will work closely with the applicant to expedite the application within a 30-day period.

For assistance, contact:
Michael Hughes
Economic Development Division Manager
(770) 528-2018
econdev@cobbcounty.org

Cobb County
ECONOMIC DEVELOPMENT

~*Applicant Information*~

Business Owner Name BRED CO., LLC (TO BE ASSIGNED TO A SPE AFFILIATE OWNED BY BRAVES DEVELOPMENT COMPANY, LLC)

Business Name BRAVES DEVELOPMENT COMPANY, LLC

Mailing Address 755 BATTERY AVE ATLANTA, GA 30339

Street Address SAME

Telephone 404 614 2191 Fax

Email MIKE.PLANT@braves.com

Applicant's Representative for Contact Regarding Request:

Name and Title MIKE PLANT, PRESIDENT + CEO

Mailing Address SAME AS ABOVE

Street Address SAME AS ABOVE

Telephone 404 614 2191 Fax

Email MIKE.PLANT@braves.com

~*Application Certification*~

I certify that the information contained in this application is accurate and complete to the best of my knowledge. I further certify that ALL of the above listed items in the instructions section have been addressed and included in this application packet. I further certify that the project has not begun and that no business transactions for this project have been executed. I further certify that the applicant does not owe any outstanding taxes or fees to Cobb County.

Signature of Owner [signature] Date

Cobb County Economic Development
P.O. Box 649, Marietta, GA 30061-0649
TEL (770) 528-2018 ~ FAX (770) 528-2126 ~ econdev@cobbcounty.org ~ www.cobbcounty.org/econdev

☐ *Special Economic Impact Incentive*

Eligibility: *In order to qualify for the special economic impact incentive program a business must be part of one of the following:*

1) *Headquarters: corporate, divisional, and/or regional;*
2) *Financial, insurance, and professional services (law, accounting, and other professional services that predominantly serve the Metropolitan Atlanta Region are not eligible);*
3) *Transportation/distribution (logistics);*
4) *Manufacturing; and/or*
5) *Emerging technologies/industries.*

Unless otherwise provided, no incentive shall be offered or made available to an eligible business for the special economic impact incentive program unless two of the following criteria are met:

1) *Add at least 150 new jobs;*
2) *Pay an average salary at least 1.25 times the county average for that industry as determined by the Georgia Department of Labor in the Employment and Wages Annual Report; and*
3) *Invests $30,000,000 or more in the county.*

Please Answer/Submit the Required Information for the Special Economic Impact Incentive:

A. Please list which of the above business types applies to your business? SEE ATTACHED
B. How many new jobs has your company created? SEE ATTACHED
C. Please provide documentation to support the number of jobs your company created.
D. What is the average salary you pay your employees by industry type? Please provide documentation by industry type. SEE ATTACHED
E. How much is your anticipated investment into the county? SEE ATTACHED
F. Please provide documentation to support your anticipated investment.

Cobb County Economic Development
P.O. Box 649, Marietta, GA 30061-0649
TEL (770) 528-2018 ~ FAX (770) 528-2126 ~ econdev@cobbcounty.org ~ www.cobbcounty.org/econdev

SUPPLEMENT TO SPECIAL ECONOMIC IMPACT INCENTIVE APPLICATION

Please note that BRED Co., LLC ("Bred"), is the current property owner, and will be transferring the subject property to a single purpose entity wholly owned by Braves Development Company, LLC, an affiliate of Bred ("Property Owner"), to develop the subject property as owner and real estate developer. The subject property will be developed to include an approximately 290,000 rentable square foot office building and related infrastructure and improvements for Project Dashboard, which is the project for which the incentives are sought. As a result, the company "Dashboard" is relocating its headquarters to the subject property, which will significantly impact the region by providing hundreds of new full time permanent jobs (thousands of construction related jobs) and investment in the County (the "Project Dashboard Impact"). More particular numbers regarding the Project Dashboard Impact are found in Dashboard's companion application for the project and attached hereto.

Property Owner:

BRED Co., LLC, a Georgia limited liability company

By:______________________________

Mike Plant, Executive Vice President

Braves Development Company, LLC, a Delaware limited liability company

By:______________________________

Mike Plant, President & CEO

Workforce Data

Headcount & Payroll Projections

Project Dashboard

* These projections reflect the Company's best estimates as of today. Future business and economic conditions may change the rate of job growth and average salaries.

Average Salary	2019	2020	2021	2022
Accounting	$69,967	$71,367	$72,744	$74,250
Administrative Assistant	$53,275	$54,341	$55,428	$56,536
Administrator	$46,459	$47,385	$48,332	$49,299
Analyst - CRM	$59,466	$60,655	$61,868	$63,106
Analyst - IT	$111,431	$113,664	$115,937	$118,256
Analyst IT INFRA	$101,412	$103,440	$105,509	$107,619
Analyst - IT Programmer	$124,265	$126,751	$129,286	$131,871
Analyst	$86,110	$87,832	$89,589	$91,380
Buyer	$54,947	$56,041	$57,162	$58,305
Chief Engineer	$201,897	$205,935	$210,054	$214,255
Clerk	$37,390	$38,138	$38,901	$39,679
Collector	$48,011	$48,976	$49,949	$50,948
Controller	$151,905	$154,943	$158,042	$161,203
Coordinator	$56,586	$57,718	$58,872	$60,050
Customer Service Representative	$50,720	$51,734	$52,774	$53,829
Designer Drafter	$58,246	$59,411	$60,599	$61,811
Engineer - Field	$121,560	$123,996	$126,478	$129,007
Engineer - IT	$127,740	$130,294	$132,900	$135,558
Engineer - Network	$77,484	$79,013	$80,614	$82,226
Engineer	$109,465	$111,655	[illegible]	$116,166
Functional Directors	$204,000	$208,080	$212,242	$216,486
Graphic Designer	$77,475	$79,024	[illegible]	$82,217
Legal	$204,000	$208,080	$212,242	$216,486
Manager	$123,994	$126,479	$129,008	$131,588
Manager Corp	$104,037	$106,111	$108,236	$110,401
Manager IT	$129,714	$132,308	$134,954	$137,654
Manager OPS	$164,203	$167,491	$170,841	$174,258
Manager POS	$204,000	$208,080	$212,242	$216,486
Operations VP	$204,000	$208,080	$212,242	$216,486
Payroll ADMIN	$68,636	$70,002	$71,402	$72,830
Planner	$78,178	$79,741	$81,336	$82,961
Recruiter	$80,784	$82,406	$84,048	$85,729
Regional Director	$204,000	$208,080	$212,242	$216,486
Regional Manager	$94,176	$96,060	$97,981	$99,940
SAP Functional SME	$127,740	$130,294	$132,900	$135,558
SR Copywriter	$75,174	$76,678	$78,211	$79,775
Supervisor	$88,433	$90,204	$92,008	$93,848
Sustainale Design Manager	$97,160	$99,103	$101,085	$103,107
Technical Writer	$77,823	$79,380	$80,967	$82,587
Technician	$62,733	$63,992	$65,272	$66,577
Technician Contract	$68,636	$70,002	$71,402	$72,830
Trainer	$79,655	$81,248	$82,873	$84,530
Trainer Specialist	$87,343	$89,092	$90,874	$92,692
US Fild VP	$204,000	$208,080	$212,342	$216,486
Web Developer	$113,902	$116,180	$118,504	$120,874
Overall Average Salary	**$67,000**	**$73,000**	**$78,000**	**$94,000**

Workforce Data

Total Taxable Payroll (cumulative)[1]	2019	2020	2021	2022
Accountant	[illegible]	[illegible]	[illegible]	$450,000
Administrative Assistant	$50,000	$820,000	$880,000	$1,070,000
Administrator	$2,740,000	$4,790,000	$4,880,000	$5,370,000
Analyst - CRM	$0	$0	$0	[illegible]
Analyst - IT	$0	$0	$120,000	$120,000
Analyst IT INFRA	$0	$0	$0	[illegible]
Analyst - IT Programmer	$0	$0	$260,000	$1,320,000
Analyst	$430,000	$880,000	$1,080,000	$2,560,000
Buyer	$490,000	$500,000	$510,000	$520,000
Chief Engineer	$0	$410,000	$420,000	$430,000
Clerk	$0	$800,000	$820,000	$910,000
Collector	$0	$4,110,000	$4,200,000	$4,280,000
Controller	$0	$310,000	$470,000	$970,000
Coordinator	$110,000	$170,000	$180,000	$840,000
Customer Service Representative	$0	$0	$0	$50,000
Designer/Drafter	$0	$480,000	$480,000	$1,240,000
Engineer - Field	$0	$0	$0	$4,390,000
Engineer - IT	$0	$0	$0	$270,000
Engineer - Network	$0	$0	$80,000	$330,000
Engineer	$0	$6,390,000	$6,720,000	$8,020,000
Functional Director	$410,000	$830,000	$1,060,000	$4,760,000
Graphic Designer	$0	$0	$0	$330,000
Legal	$0	$0	$0	$430,000
Manager	$1,120,000	$3,790,000	$3,870,000	$9,080,000
Manager Corp	$210,000	$320,000	$320,000	$550,000
Manager IT	$0	$0	$130,000	$550,000
Manager OPS	$160,000	$170,000	$170,000	$1,050,000
Manager POS	$0	$210,000	$210,000	$430,000
Operations VP	$200,000	$1,040,000	$1,050,000	$3,460,000
Payroll ADMIN	$0	$70,000	$70,000	$70,000
Planner	$80,000	$80,000	$80,000	$80,000
Recruiter	$0	$0	$0	$170,000
Regional Director	$0	$0	$0	$220,000
Regional Manager	$190,000	$290,000	$290,000	$300,000
SAP Functional SME	$0	$0	$400,000	$410,000
SR Supervisor	$0	$0	$0	$80,000
Supervisor	$350,000	$2,260,000	$2,300,000	$2,630,000
Sustainable Design Manager	$0	$0	$0	$100,000
Technical Writer	$0	$0	$0	$330,000
Technician	$0	$130,000	$130,000	$1,200,000
Technician Contract	$0	$0	$0	$150,000
Trainer	$0	$0	$0	$420,000
Trainer Specialist	$0	$0	$0	$190,000
US Fld VP	$0	$210,000	$210,000	$650,000
Web Developer	$0	$0	$0	$360,000
Total Payroll	**$6,610,000**	**$29,470,000**	**$31,570,000**	**$61,950,000**

Economic and Fiscal Impact

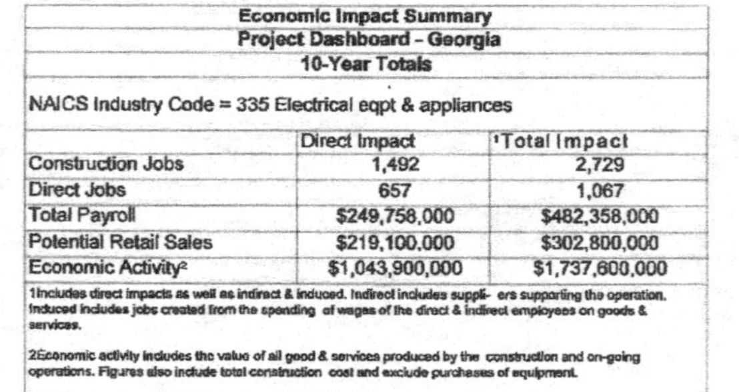

Economic Impact Summary		
Project Dashboard - Georgia		
10-Year Totals		
NAICS Industry Code = 335 Electrical eqpt & appliances		
	Direct Impact	[1]Total Impact
Construction Jobs	1,492	2,729
Direct Jobs	657	1,067
Total Payroll	$249,758,000	$482,358,000
Potential Retail Sales	$219,100,000	$302,800,000
Economic Activity[2]	$1,043,900,000	$1,737,600,000

1Includes direct impacts as well as indirect & induced. Indirect includes suppli- ers supporting the operation. Induced includes jobs created from the spending of wages of the direct & indirect employees on goods & services.

2Economic activity includes the value of all good & services produced by the construction and on-going operations. Figures also include total construction cost and exclude purchases of equipment.

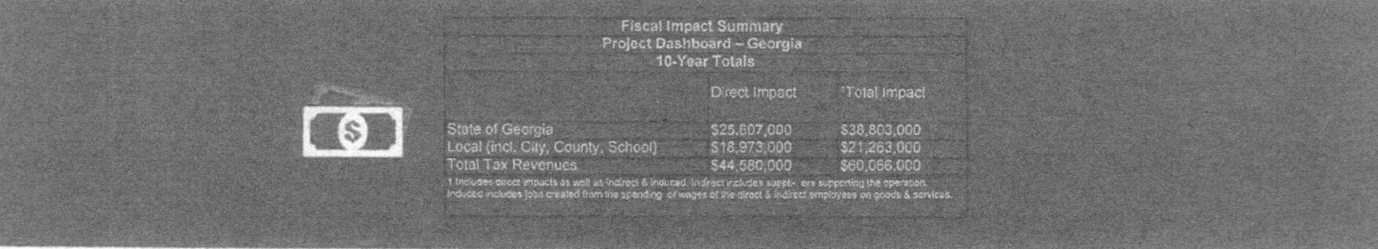

Fiscal Impact Summary		
Project Dashboard – Georgia		
10-Year Totals		
	Direct Impact	[1]Total Impact
State of Georgia	$25,607,000	$38,803,000
Local (incl. City, County, School)	$18,973,000	$21,263,000
Total Tax Revenues	$44,580,000	$60,066,000

1 Includes direct impacts as well as indirect & induced. Indirect includes suppli- ers supporting the operation. Induced includes jobs created from the spending of wages of the direct & indirect employees on goods & services.

Appendix B

Proposed Request for Interest (RFI)

Request for Interest *(Proposed)*

1. PURPOSE

The Department of the Navy (DON) is requesting letters of interest (LOI) from parties who are interested in using Navy property situated at Naval Air Station Oceana, Virginia Beach, Virginia. The LOIs will be used to engage in preliminary negotiations with the DON for a P3 partnership on developing six parcels of perimeter real estate as included in Exhibit A. The DON will issue a formal solicitation to offer the Navy property for commercial use by interested parties.

LOCATION OF PROPERTIES

Naval Air Station, Oceana
Parcels 1–6, Exhibit A
Virginia Beach, VA

DEPARTMENT OR AGENCY

Department of the Navy (DON)
By: Naval Facilities Engineering Command, Mid-Atlantic: AM11-PM
9324 Virginia Avenue
Norfolk, VA 23511

2. DESCRIPTION

#	Parcel	Acres Total	Acres Wetlands	Acres Usable
1	Commissary West	29.9	4.2	25.7
2	Commissary East	24.5	1.1	23.3
3	Harpers Rd / AG Field	19.5	0.6	18.8
4	Harpers Rd / Old Housing	55.2	0.0	55.2
5	Oceana Blvd / AG Field	200.4	110.0	90.4
6	Owl's Creek	187.9	100.0	87.9
	Total	517.3	216.0	301.3

Six parcels as shown in Exhibit A are located at Naval Air Station, Oceana, Virginia Beach, Virginia. The gross area of the parcel is approximately shown as above.

3. ENCUMBRANCES/ENVIRONMENTAL CONSTRAINTS

An Environmental Checklist is attached to this RFI as Exhibit B. It contains pertinent information regarding the known and unknown conditions of the Parcel to potentially be out-leased. The Environmental Checklist does not represent all known environmental conditions. The anticipated permitting requirements for each parcel have been examined and included.

4. PROJECT VISION

An initial market analysis and site analysis are included in Exhibit C. The vision for each site is not restricted by its envisioned use, only by potential uses and encumbrances listed below. Initial visions for the site are mixed use dense development along Oceana Boulevard following municipal guidelines for Virginia City's comprehensive plan. Programming amounts are to be determined by developer proposal in conjunction with the Government via a P3 partnership, but are initially envisioned as office, commercial, retail, healthcare, and multi-family mixed use with a potential option for government leased warehousing and multifamily. A phased approach to development is envisioned with only 10 percent of available acreage released per year pending P3 negotiations and further study of market conditions. Out-leasing options are initially envisioned as fifty-year long term leasing in exchange for in-kind consideration and cash options.

5. POTENTIAL USES

a. This parcel may not be used for any of the following activities:

1. Residential, including single family, multi-family, temporary, or permanent
2. Daycares, child education centers, day nurseries
3. Public or private education
4. Churches, funeral homes, chapels
5. Recreational campgrounds, auditoriums, assembly halls, sports arenas
6. Lodges for organizations, private clubs
7. Airports, heliports, helistops
8. Movie theaters, drive in theaters, indoor recreational, and amusement facilities
9. Athletic clubs
10. Group homes
11. Hospitals, nursing homes
12. Spectator sports
13. Manufacturing, storage, or handling of explosives, and petrochemicals, except for petrochemicals accessory to any use that is not prohibited
14. Shopping centers
15. Retail sales in locations outside relocated base perimeter and no more than 10 percent of usable building space within the base perimeter
16. Offices or office buildings

b. Because of proximity to NAS Oceana's airfield, any proposed structure will likely require submission of an Obstruction Evaluation/Airport Airspace Analysis (OE/AAA) to the Federal Aviation Administration (FAA). Use of raised construction cranes will also likely require an OE/AAA submission. More information is available at https://oeaaa.faa.gov/oeaaa/external/portal.jsp or by contacting the NAS Oceana Planning Liaison at OceanaComments@navy.mil.
c. Base perimeter fencing will be relocated by the government via in-kind consideration prior to project out-lease. Fire and emergency services as well as infrastructure maintenance will be provided by the government in exchange for leasing into an area with no municipal tax or business licenses, pending negotiation details.
d. Thirty days are provided for response. Questions may be submitted prior to the deadline and will be collected no later than 7 days prior to the due date and answers released to all respondees via amendment to this RFI no later than 72 hours after deadline. Responses will not be ranked or evaluated, however interested parties will be contacted via the below listed Contracting Officer for potential P3 partnership engagement.

RFI Proposed Update Explained

The RFI is a request for interest, different from a "construction RFI" or request for information which has the same acronym. The request for information's intent is different, which is to advertise interest in partnering for a project in the initial development phase prior to any formal solicitation and identify potential partners for further negotiation and interest. No proposal is developed by an interested party at this phase. Although it is a general interest statement, it is important to let potential partners know the government's intentions and flexibility as well as key parcel data to engage private partners for developing an EUL agreement.

Although the government has much leeway with 517 acres of potential leasable land within close proximity of the area, the market analysis indicated a phased approach is more appropriate as the market could not quickly absorb such a large amount of real estate. Based on the site analysis, the Owl's Creek property is much less desirable for development as 40 percent of the acreage is wetland, the environmental impact study identified a rare migratory bird which may limit development due to the Environmental Species Act, and the majority of land touching existing utilities and infrastructure is non-buildable due to wetland locations.

Thus, Owl Creek would be a good target to fund improvements to make other parcels more desirable via in-kind consideration from a partner who would desire such land. The municipality has a stated goal of greenspace preservation, desire for open space, and land near the resort area, all of which is located on Owl Creek. Prior to an RFI, engagement with the local municipality could be warranted to potentially exchange acreage of the parcel for the construction cost of relocating all perimeter military fencing all the other parcels in order to complete site preparation, roughly \$2.5 to \$3.5 million dollars. This would make each site much more desirable as it would eliminate lengthy timelines for security perimeter relocations while disposing of land to the local municipality that is largely a maintenance cost but not a heavily utilized area for the government now or in the foreseeable future. The RFI includes phased language and would exclude Owl's Creek if the local municipality were interested in a land exchange for in-kind consideration.

One area specifically left out of the RFI is the amount of programming envisioned from the market analysis. Although REIS software was used, the government is largely bad at estimating marketability and any included estimates are more to give a contractor a rough idea rather than specific requirements for development. Instead, a project vision section is included in the proposed RFI to give the developer a ballpark idea for the property vision the Navy would find more desirable for P3 partnership. Further, the initial market analysis is included as an enclosure to provide better initial information to generate interest. Most developers will quickly be able to generate more robust products, however the RFI is to initiate initial ideas and partnerships and is not designed to encumber the private sector with expenses prior to engagement.

The RFI combines all six parcels (or five if Owl Creek is negotiated away) into one RFI rather than six separately posted documents. This streamlines communication and allows for developers to target portions of parcels on both sides of Oceana Boulevard. It also reduces confusion from a federal RFI posting system. The RFI includes market and site analysis, and recommended base desires for either in-kind consideration are already predetermined from the MEP examination.

Glossary of Key P3 Terms

Adjustable-Rate Mortgage (ARM) A mortgage loan with an interest rate subject to change over the term of the loan. The interest rate is tied to the performance of a specified market rate.

Amortization The act of paying down principal over its term. In a typical mortgage loan, the principal is scheduled to be paid off or fully amortized over the loan term.

Average Hourly Earnings A monthly reading by the Bureau of Labor Statistics of the earnings of hourly plant and non-supervisory workers in the private sector.

Basis Point A basis point is one one-hundredth of a percentage point. For example, if mortgage rates fall from 7.50 percent to 7.47 percent, they have declined three basis points. A full percentage point is 100 basis points.

Brownfield With specific legal exclusions and additions, the term "brownfield site" means real property, the expansion, redevelopment, or reuse, which may be complicated by the presence, or potential presence, of a hazardous substance, pollutant, or contaminant. Cleaning up and reinvesting in these properties protects the environment, reduces blight, and takes development pressures off greenspaces and working lands.

Build/Operating Agreements These include: Build-Operate Transfer (BOT); Build-Own-Operate-Transfer (BOOT); Design-Build-Finance-Operate (DBFO); Design-Construct-Manage-Finance (DCMF); Independent Power Producer (IPP); Build-Own-Operate (BOO).

Business Improvement District (BID) A privately directed and publicly sanctioned organization that supplements public services within geographically defined boundaries by generating multiyear revenue through a compulsory assessment on local property owners and/or businesses.

Cash-Out Refi Refinancing a mortgage in which the new principal (the borrowed amount) exceeds the original loan's outstanding principal by at least 5 percent. In other words, the homeowner is taking equity out of the home.

Census Tract An area delineated by the U.S. Bureau of the Census for which statistics are published; in urban areas, census tracts correspond roughly to neighborhoods.

Central Business District (CBD) Office buildings located in the central business district are in the heart of a city. In larger cities like Chicago or New York, and some medium-sized cities like Orlando or Jacksonville, these buildings would include high-rises in downtown areas. Other office buildings can be found in other areas of the metropolitan area, including suburban office buildings. This office space classification generally includes mid-rise structures of 80,000–400,000 square feet located outside of a city center. Cities will also often have suburban office parks that assemble several different mid-rise buildings into a campus-like setting.

Cityscape The urban equivalent of a landscape.

Co-housing Projects The intentional community of private homes clustered around shared space. Each attached or single-family home has standard amenities, including a private kitchen. Shared space usually features a typical house, including a large kitchen and dining area, laundry, and recreational spaces. Shared outdoor space may include parking, walkways, open space, and gardens. Neighbors also share resources like tools and lawnmowers.

Commercial Development Commercial property (also called commercial real estate, investment, or income property) refers to buildings or land intended to generate a profit, either from capital gain or rental income. Commercial property includes office buildings, industrial property, medical centers, hotels, malls, retail stores, farmland, multifamily housing buildings, warehouses, and garages. In many states, residential property containing more than a certain number of units qualifies as a commercial property for borrowing and tax purposes.

Commercialization Transforming an area of a city into a place attractive to residents and tourists alike in terms of economic activity.

Community Redevelopment Agency (CRA) A public entity created by a city or county to implement the community redevelopment activities outlined under the Community Redevelopment Act which was enacted in 1969 (Chapter 163, Part III, Florida Statutes).

Concurrency A shorthand expression for a set of land use regulations that local governments are required (by the Florida Legislature) to adopt to ensure that new development does not outstrip local government's ability to handle it. For a real estate development to "be concurrent" or "meet concurrency," the local government must have enough infrastructure capacity to serve each proposed real estate development. Specifically, concurrency regulations require that the local government has stormwater, parks, solid waste, water, sewer, and mass transit facilities to serve each proposed development. Together, these seven public services are known as "concurrency facilities."

Conforming Mortgage Loan Any mortgage loan at or below the amount Fannie Mae and Freddie Mac can purchase and/or securitize in the secondary mortgage market.

Construction Loan A temporary loan used to pay for the building of a house.

Consumer Confidence Index A measure of confidence households have in the economy. The index is released monthly by the Conference Board.

Consumer Price Index (CPI) The CPI is the measurement of the average change in prices paid by consumers for a fixed market basket of a wide variety of goods and services to determine inflation's underlying rate. The broadest and most quoted CPI figure reflects the average change in the prices paid by urban consumers (about 80 percent of the U.S. population). The so-called "core CPI" excludes the volatile food and energy sectors.

Conventional Mortgage Loan A conventional mortgage loan is any mortgage loan not guaranteed or insured by the government (typically through FHA or V.A. programs).

Corporate Financing A company borrows money against its proven credit position and ongoing business and invests in the project.

Credit Report An individual's report of borrowing and repayment history.

Credit Score A three-digit number based on an individual's credit report used to indicate credit risk.

Edge City A large node of office and retail active activities on an urban area's edge.

Eminent Domain (ED) The right of a government or its agent to expropriate private property for public use, with payment of compensation. *Public Use*: Traditional use of ED to secure land for utilities, schools, roads etc. *Public Purpose*: The public can have access to and the ability to use an area such as a town center or a park. *Public Benefit*: Post *Kelo v. New London Conn.* Eminent domain can be used for the greater good of the public even though it may not impact the public directly, such as using ED for private development to spur economic growth.

Employment (Payroll) The number of non-farm employees on the payrolls of more than 500 private and public industries, issued monthly by the Bureau of Labor Statistics.

Employment Cost Index A quarterly index used to gauge the change in civilian labor cost, including salaried workers.

Existing-Home Sales Based on the number of closings during a particular month. Because of the one- to two-month period between a signed purchase contract and a closing, existing home sales are more influenced by mortgage rates a month or two earlier than the prevailing mortgage rate during the month of closing.

Fannie Mae and Freddie Mac The United States' two federally chartered and stockholder-owned mortgage finance companies. Forbidden by their charters from originating loans (that is, from providing mortgage loans on a retail basis), these two Government-Sponsored Enterprises (GSEs) purchase and/or securitize mortgage loans made by others. Due to their directive to serve low-, moderate-, and middle-income families, the GSEs have loan limits on the purchase or securitization of mortgages.

Federal Funds Rate The rate banks charge each other on overnight loans made between them. These loans are generally made so banks can cover their daily cash flow and reserve requirements. The federal government doesn't actually set the fed funds rate, which is determined by the funds' supply and demand.

Instead, it forms a target rate and affects the supply of funds through its securities sales purchases.

Federal Open Market Committee (FOMC) The Federal Reserve arm that sets monetary policy, the FOMC, is scheduled to meet eight times a year. The 12 members of the FOMC include the seven governors of the Federal Reserve System, the president of the New York Federal Reserve Bank, and, on a rotating basis, four of the presidents from 11 other regional Federal Reserve Banks.

Fixed-Rate Mortgage (FRM) A mortgage loan with an interest rate that does not change over the loan term.

Gentrification This is a controversial practice. Gentrification is converting an urban neighborhood from a predominantly low-income renter-occupied area to a mostly middle-class owner-occupied area. The renewal and rebuilding process accompanying the influx of middle-class or affluent people into deteriorating areas often displaces lower-income residents.

Government Financing The government borrows money and provides it to the project through on-lending, grants, or subsidies or where it provides guarantees of indebtedness.

Greenbelt A ring of land maintained as parks, agriculture, or other open space types limits the sprawl of an urban area.

Greenfield A greenfield project is one that lacks constraints imposed by prior work. The analogy is to construction on greenfield land where there is no need to work within existing buildings or infrastructure constraints.

Gross Domestic Product (GDP) The value of all the final goods and services produced in the United States over a particular period. Available quarterly from the Bureau of Economic Analysis.

Home Equity Home equity is the difference between the house's current value and the amount of money owed on the mortgage.

Home Equity Line of Credit A home equity line of credit is an open credit line secured by the equity in your home.

Home Improvement Loan Money lent to a property owner for home repairs and remodeling.

Homeownership Rate The number of households residing in their own home divided by the total number of households in the United States. The U.S. Census Bureau releases an estimate of the homeownership rate based on a quarterly survey.

Hotels, Extended Stay These hotels have larger rooms, small kitchens, and are designed for people staying a week or more.

Hotels, Full-Service Full-service hotels are usually located in central business districts or tourist areas and include the big-name flags like Four Seasons, Marriott, and Ritz Carlton.

Hotels, Limited-Service Hotels in the limited-service category are usually boutique properties. These hotels are smaller and don't typically provide amenities such as room service, on-site restaurants, or convention space.

House Price Index (HPI) A quarterly measure of the change in single-family house prices released by the Office of Federal Housing Enterprise Oversight. The HPI is a repeat sales index, meaning it measures average price changes in repeat sales or refinancing on the same properties. It is based on mortgages purchased or securitized by Fannie Mae and Freddie Mac. Homes with mortgages above the Fannie/Freddie conforming loan limit, and houses insured or guaranteed by the FHA, VA, or other federal government entity, are not included in the sampling.

Housing Starts The Census Bureau's monthly count of the number of private residential structures on which construction has started or permits have been issued.

Industrial: Heavy Manufacturing This category of industrial property is a particular use category for most large manufacturers. These properties are heavily customized with machinery for the end-user and usually require substantial renovation to re-purpose for another tenant.

Industrial: Light Assembly These structures are much simpler than the above heavy manufacturing properties and usually can be easily reconfigured. Typical uses include storage, product assembly, and office space.

Industrial: Flex Warehouse Flex space is industrial property that can be easily converted and typically includes a mix of industrial and office space.

Industrial: Bulk Warehouse These properties are extensive, generally in the range of 50,000–1,000,000 square feet. These properties are often used for regional distribution of products and require easy access by trucks entering and exiting highway systems.

Infill Building on empty parcels of land within a checkerboard pattern of development.

Informal Sector Economic activities that take place beyond official record, not subject to formalized systems of regulations or remuneration.

Infrastructure The underlying framework of services and amenities needed to facilitate productive activity.

Interest Rate A measure of the cost of borrowing.

Jumbo Mortgage Loan A mortgage loan for an amount exceeding the Fannie Mae and Freddie Mac loan limit. Because the two agencies can't purchase the lender's loan, jumbo loans carry higher interest rates.

Lateral Commute Traveling from one suburb to another and going from home to work.

Lift Station Wastewater lift stations are facilities designed to move wastewater from lower to higher elevation, mainly where the source's height is not sufficient for gravity flow and/or when the use of gravity conveyance will result in excessive excavation depths and high sewer construction costs.

Loan-To-Value Ratio (LTV) In a mortgage loan, the amount borrowed relative to the value of the property. An LTV of 80 percent means the mortgage loan is 80 percent of the property's value, with the borrower making a 20 percent down payment.

Low Income Housing Tax Credit Subsidizes the acquisition, construction, and rehabilitation of affordable rental housing for low- and moderate-income tenants.

Mean Home Price (of New or Existing Homes Sold) The mean home price is a mathematical average of the costs of all homes sold in the period, typically monthly. The mean price of homes sold generally runs higher than the median price due to the number of very high-priced homes.

Median Home Price (of New or Existing Homes Sold) The median home price is the median price of all the homes sold within 30 days. Median home prices are generally a better indicator of home price trends than average home prices.

Megacities Cities with more than 10 million people.

Megalopolis The continuous urban complex in the Northeastern United States.

Metropolitan Area Within the United States, an urban area consisting of one or more whole country units, usually containing several urbanized areas or suburbs, which all act together as a coherent economic whole.

Metropolitan Statistical Area (MSA) In the United States, an urbanized area of at least 50,000 population, the country within which the city is located and adjacent countries meeting one of several tests indicating a functional connection to the central city.

Micropolitan Statistical Area An urbanized area of between 10,000 and 50,000 inhabitants, the country in which it is found, and adjacent countries tied to the city.

Mixed-Income Mixed-income housing may include housing that is priced based on the dominant housing market (market-rate units), with only a few units priced for lower-income residents. It may not have any market-rate units and be built exclusively for low- and moderate-income residents.

Mixed-Use In a broad sense, mixed-use development—any urban, suburban, or village development, or even a single building, blends a combination of residential, commercial, cultural, institutional, or industrial uses. Those functions are physically and functionally integrated, and provide pedestrian connections.

Mortgage A loan to buy real estate and secured by the real estate.

Mortgage Application Index (Purchase) An index published weekly by the Mortgage Bankers Association of America to gauge the number of applications submitted to purchase a home. The survey covers about 40 percent of all retail and residential mortgage transactions.

Mortgage Application Index (Refinance) A mortgage application index (refinance) is an index published weekly by the Mortgage Bankers Association of America to gauge the number of applications submitted for a home's refinancing. The survey covers about 40 percent of all retail and residential mortgage transactions.

Mortgage Broker A person or company that acts as a mediator between borrowers and lenders.

Multi-family Housing Multi-family residential (also known as a multi-dwelling unit or MDU) is a classification of housing whereby multiple separate housing units for residential inhabitants are contained within one building or several buildings within one complex. A common form is an apartment building. Sometimes, units in a multi-family residential building are condominiums, where the units are typically owned individually rather than leased from a single apartment building owner. Many intentional communities incorporate multi-family residences, such as in co-housing projects.

Multi-family Housing: Garden Apartments Suburban garden apartments started popping up in the 1960s and 1970s, as families moved from urban centers to the suburbs. Garden apartments are typically 3–4 stories with 50–400 units, no elevators, and surface parking.

Multi-family Housing: Mid-rise Apartments These properties are usually 5–9 stories, with between 30 and 110 apartment units and elevator service. These are often constructed in urban infill locations.

Multi-family Housing: High-rise Apartments High-rise apartments are found in larger markets, usually have 100+ units, and are professionally managed.

Multiplier Effect The direct, indirect, and induced consequences of change in an activity; in urban geography, the expected addition of non-basic workers and dependents to a city's local employment and a population that accompanies new primary sector employment.

New Markets Tax Credit (NMTC) The NMTC Program attracts private capital into low-income communities by permitting individual and corporate investors to receive a tax credit against their federal income tax in exchange for making equity investments in specialized financial intermediaries called Community Development Entities (CDEs).

Office Classification Office buildings are usually loosely grouped into one of three categories: Class A, Class B, or Class C. These classifications are all relative and largely depend on context. Class A buildings are considered the best of the best in terms of construction and location. Class B properties might have high-quality construction but with a less desirable location. Class C is everything else.

Office Park An area where several office buildings are built together on landscaped grounds.

Opportunity Zones An economic development tool that allows people to invest in distressed areas in the United States. Their purpose is to spur economic growth and job creation in low-income communities while providing tax benefits to investors.

Planned Communities A planned community is a city, town, or community designed from scratch and developed to follow the plan.

Program-Related Investments (PRIs) PRIs hold incredible potential for the social enterprise arena. Rather than giving away money through grants, PRIs allow foundations to make investments as loans or equity stakes in the hopes of regaining their assets plus a reasonable rate of return.

Producer Price Index (PPI) The PPI measures the average change in the selling prices of goods and services sold by domestic producers and is an inflation indicator released monthly by the Bureau of Labor Statistics.

Project Financing Where non- or limited-recourse loans are made directly to a special-purpose vehicle.

Racial Steering The practice in which real estate brokers guide prospective home buyers towards or away from specific neighborhoods based on their race.

Redlining A process by which banks draw lines on a map and refuse to lend money to purchase or improve the property within the boundaries.

Real Estate Owned (REO) A term used in the United States to describe a class of property owned by a lender—typically a bank, government agency, or government loan insurer—after an unsuccessful sale at a foreclosure auction.

Restrictive Covenants Statements written into a property deed that restrict the use of land in some way.

Retail: Strip Center Strip centers are smaller retail properties that may or may not contain anchor tenants. An anchor tenant is simply a larger retail tenant, which usually draws customers into the property. Examples of anchor tenants are Walmart, Publix, and Home Depot. Strip centers typically contain a mix of small retail stores like Chinese restaurants, dry cleaners, nail salons, etc.

Retail: Community Retail Center Community retail centers are generally in the range of 150,000–350,000 square feet. Multiple anchors occupy community centers, such as grocery stores and drug stores. Additionally, it is common to find one or more restaurants located in a community retail center.

Retail: Power Center A power center generally has several smaller, inline retail stores but is distinguished by a few major box retailers, such as Walmart, Lowes, Staples, Best Buy, etc. Each big-box retailer usually occupies between 30,000 and 200,000 square feet, and these retail centers typically contain several out parcels.

Retail: Regional Mall Malls range from 400,000 to 2,000,000 square feet and generally have a handful of anchor tenants such as department stores or big-box retailers like Barnes & Noble or Best Buy.

Retail: Out Parcel Most larger retail centers contain one or more out parcels, which are parcels of land set aside for individual tenants such as fast-food restaurants or banks.

Retention/Detention Pond A detention pond is a low-lying area designed to temporarily hold a set amount of water while slowly draining to another location. These ponds control floods when large amounts of rain could cause flash flooding. A retention pond is designed to hold a specific amount of water indefinitely.

Second Mortgage A real estate mortgage that has already been pledged as collateral against another mortgage. Typically used to draw cash from home for other purposes.

Securitization The pooling of mortgage loans into a mortgage-backed security. The principal and interest payments from the individual mortgages are paid out to the MBS security holders.

Short Sale A sale of real estate in which the net proceeds from selling the property will fall short of the debts secured by liens against the property. In this case, if all lien holders agree to accept less than the amount owed on the debt, the sale of the property can be accomplished.

Single-family Housing A single-family (home, house, or dwelling) means that the building is usually occupied by just one household or family and consists of only one dwelling unit or suite. In some jurisdictions, allowances are made for basement suites or mother-in-law suites without changing the "single family" description. However, it does exclude any short-term accommodation (hotels, motels, inns) and large-scale rental accommodation (rooming or boarding houses, apartments).

Smart Growth Legislation and regulations to limit suburban sprawl and preserve farmland.

Sprawl Sprawl is developing new housing sites with relatively low density at locations that are not contiguous to the existing built-up area.

Suburbanization The movement of upper- and middle-class people from urban core areas to the surrounding outskirts to escape pollution as well as deteriorating social conditions.

Tax Increment Financing (TIF) A value capture revenue tool that uses taxes on future gains in real estate values to pay for new infrastructure improvements. TIFs are authorized by state law in nearly all 50 U.S. states and begin with the designation of a geographic area as a TIF district.

Topography The topography is a detailed map of the surface features of the land. It includes the mountains, hills, creeks, and other bumps and lumps on a particular hunk of earth.

Underwriting The determination of the risk a lender would assume if a particular mortgage loan application is approved.

Urban Growth Rate The rate of growth of an urban population.

Urban Morphology The form and structure of cities, including street patterns and the size and shape of buildings.

Urbanized Area The city as art.

Zoning Dividing an area into zones or sections reserved for different purposes such as residence and business and manufacturing.

References

Chapter 1

Corrigan, M. B., Hambene, J., Hudnut III, W., Levitt, R. L., Stainback, J., Ward, R., & Witenstein, N. (2005). *Ten Principles for Successful Public/Private Partnerships*. Washington DC: Urban Land Institute.

Emrath, P., & Sugrue Walter, C. (2022). *Regulation: 40.6 Percent of the Cost of Multifamily Development*. Report for NAHB and NMHC.

Friedman, S. B. (Ed.) (2016). *Successful Public/Private Partnerships: From Principles to Practices*. Washington DC: Urban Land Institute.

Nelson, A. C. (2014). *Foundations of Real Estate Development Finance: A Guide to Public-Private Partnerships*. Washington DC: Island Press.

Shaver, L. (2022, June 16). Report: Regulations Account for 40% of Development Costs. In *Multifamily Dive*. Available at: https://www.multifamilydive.com/news/report-regulations-account-for-40-of-development-costs/625474/

Stainback, J. (2000). *Public/Private Finance and Development: Methodology, Deal Structuring, Developer Solicitation*. New York, NY: John Wiley and Sons.

Chapter 3

Benjamin, C. (2022, April 19). Panthers Owner David Tepper Dumps Plans for $800m South Carolina Headquarters Mid-construction, Per Report. *CBSSports.com*. Retrieved May 3, 2022, from https://www.cbssports.com/nfl/news/panthers-owner-david-tepper-dumps-plans-for-800m-south-carolina-headquarters-mid-construction-per-report/

Black's Law Dictionary Free Online Legal Dictionary 2nd ed. (2014, September 25). What Is Contract? Definition of Contract (*Black's Law Dictionary*). TheLawDictionary.org. Retrieved May 2, 2022, from https://thelawdictionary.org/contract/

Brierton, J., Shiff, B., & Hibberd, N. (2022, September 7). "Fraudulent intent": City of Rock Hill Seeks $20 Million for Abandoned Panthers Project. Wcnc.com. Retrieved October 23, 2022, from https://www.wcnc.com/article/sports/nfl/panthers/city-rock-hill-legal-panthers-david-tepper-training-camp/275-2c560826-73d5-41ed-a8b9-679d15b642ef

Bruce, M. (2021, October 14). Tallest Elevator Test Tower in U.S. the Star of Company's $200-million Cobb Headquarters. *The Atlanta Journal—Constitution*. Retrieved May 3, 2022, from https://www.ajc.com/neighborhoods/cobb/tallest-elevator-test-tower-the-star-of-companys-new-200-million-headquarters-in-the-battery/PGZHLPH775C6NCU4CMNK6CBBWA/

Cobb County, Georgia. (n.d.). Development Authority of Cobb County. Retrieved October 25, 2022, from https://www.cobbcounty.org/board/county-clerk/boards-and-authorities/development-authority-cobb-county

Cunningham, C. (2018, September 28). Cobb Confirms Incentives for Elevator Tower. *The Atlanta Journal—Constitution*. Retrieved May 3, 2022, from https://www.ajc.com/news/local/cobb-confirms-incentives-for-elevator tower/x33OFpd7GtFqdmMhvyWErJ/

Encyclopedia Britannica, inc. (n.d.). Public-Private Partnership. *Encyclopedia Britannica*. Retrieved May 2, 2022, from https://www.britannica.com/topic/public-private-partnership

Eskieva, I., & Shiff, B. (2022, April 20). Panthers Terminating Rock Hill Project Agreements. Wcnc.com. Retrieved May 3, 2022, from https://www.wcnc.com/article/sports/nfl/panthers/panthers-terminating-rock-hill-south-carolina-project-agreements/275-be9a47a6-22f4-448b-aa84-8801a29278db

Ga. Const. 1983, art. III, § VI, para.VI, as amended through January 1, 2017. Retrieved October 24, 2022, from https://www.senate.ga.gov/Documents/gaconstitution.pdf

Ga. Const. 1983, art. VII, § I, para. III, as amended through January 1, 2017. Retrieved October 24, 2022, from https://www.senate.ga.gov/Documents/gaconstitution.pdf

Wickert, D. (2021, June 19). Turning I-285 Tolls Over to Private Firm Comes with Rewards, Risks. *The Atlanta Journal—Constitution*. Retrieved May 3, 2022, from https://www.ajc.com/politics/turning-i-285-tolls-over-to-private-firm-comes-with-rewards-risks/SYUPKOSHWJBEJE22ICOJN72X4Y/

Yescombe, E. R. (2013). *Principles of Project Finance* (2nd ed.). New York: Elsevier.

Chapter 4

Corrigan, M. B. (2005). *Ten Principles for Successful Public/Private Partnerships*. Washington, DC: Urban Land Institute.

Land Design/Research Inc. (1989). Downtown Development Strategy, Greenville, South Carolina, https://www.greenvillesc.gov/DocumentCenter/View/6511/1989-Downtown-Development-Strategy-PDF?bidId=

Chapter 6

Agha, N. (2013). The Economic Impact of Stadia and Teams: The Case of Minor League Baseball. *Journal of Sports Economics*, *14*(3), 227–252.

Agha, N., & Ascher, D. A. (2016). An Explanation of Economic Impact: Why Positive Impacts Can Exist for Smaller Sports. *Sport, Business, and Management: An International Journal*, *6*(2), 182–204.

City of Greenville (2020, April 21). "City of Greenville—fluor field/field house.pdf," www.greenvillesc.gov/agendacenter

City of Greenville (2020, April 22). "Ordinance_Fluor Field Development Agreement.pdf," www.greenvillesc.gov/agendacenter

City of Greenville (2020, April 22). "Ordinance Lease in West End.pdf," www.greenvillesc.gov/agendacenter

City of Greenville (2020, April 22). "Resolution_Fluor Field 2016-36.pdf," www.greenvillesc.gov/agendacenter

City of Greenville (2020, April 21). "Resolution_Property purchase_Fluor Field.pdf," www.greenvillesc.gov/agendacenter

Connor, E. (2020, February 6). Downtown Greenville's Latest Boutique Hotel Coming to the West End. *Greenville Online*. https://www.greenvilleonline.com/story/news/2020/02/06/downtown-greenville-west-end-kimpton-hotel/4678417002/

Coomes Jr., J., & Scheuer, D. (2016). Introduction. In S. Friedman (Ed.), *Successful Public/Private Partnerships: From Principles to Practices* (pp. 2–5). Washington DC: Urban Land Institute.

Coomes Jr., J., Burkland, M., & Fullerton, J. (2016). What We Mean When We Say Public/Private Partnership. In S. Friedman (Ed.), *Successful Public/Private Partnerships: From Principles to Practices* (pp. 6–13). Washington DC: Urban Land Institute.

Corrigan, M. B., Hmabree, J., Hudnut III, W., Levitt, R. L., Stainback, J., Ward, R., & Witentstein, N. (2005). *Ten Principles for Successful Public/Private Partnerships*. Washington DC: Urban Land Institute.

Cresswell, J. W. (2003). *Research Design: Quaitative, Quantitative, and Mixed Methods Approaches* (2nd ed.). Thousand Oaks: Sage Publications.

Delmon, J. (2017). *Public-Private Partnership Projects in Infrastructure: An Essential Guide for Policy Makers.* New York: Cambridge University Press

Greenvillesc.gov. (n.d.). *History of City Managers*. https://www.greenvillesc.gov/300/History-of-City-Managers

Heather (2019, March 7). New Developments Coming to the West End. *GVL Today*. https://gvltoday.6amcity.com/new-developments-coming-west-end/

Roy, D. (2008). Impact of New Minor League Baseball Stadiums on Game Attendance. *Sports Marketing Quarterly*, *17*(3), 146–153.

Shuttleworth, M. (2008, April 1). Case Study Research Design. Retrieved September 14, 2020 from Explorable.com: https://explorable.com/case-study-research-design

Smith, E. P (2020, April 9). More than Just a Ballpark: Greenville Drive Celebrates Its 15th Anniversary. *Greenville Journal*. https://greenvillejournal.com/sports/greenville-drive-celebrates-15th-anniversary/

Stainback, J. (2000). *Public/Private Finance and Development: Methodology, Deal Structure, Developer Solicitation.* New York: John Wiley & Sons.

Van Holm, E. J. (2018). Left on Base: Minor League Baseball Stadiums and Gentrification. *Urban Affairs Review*, 54(3), 632–657.

Van Holm, E. J. (2019). Minor Stadiums, Major Effects? Patterns and Sources of Redevelopment Surrounding Minor League Baseball Stadiums. In *Urban Studies*, *56*(4), 672–688.

"West End Events at Fluor Field." (2020, April 22). https://www.visitgreenvillesc.com/listing/west-end-events-at-fluor-field/6783/.

Whitworth, N. P., & Neal, M. D. (2020, April 22). How Greenville, South Carolina, Brought Downtown Back: A Case Study in 30 Years of Successful Public/Private Collaboration, Spring 2008. http://www.saveourgateways.com/HowGreenville.php

Worthy, C. (2020, January 2). 2020 Events Calendar. *Greenville Online*. https://www.greenvilleonline.com/story/life/2020/01/02/2020-events-calendar/2796042001/

Chapter 7

Abbott, M. P. (2018). North Carolina's Research Triangle Park: A Success Story of Private Industry Fostering Public Investment to Create a Homegrown Commercial Park. *Campbell Law Review*, *40*, 569–610.

Ammon, F. R. (2016). *Bulldozer: Demolition and Clearance of the Postwar Landscape*. New Haven, CT: Yale University Press.

Audirac, I. (2018). Shrinking Cities: An Unfit Term for American Urban Policy? *Cities*, *75*(May), 12–19.

Audretsch, D. B. (2003). Innovation and Spatial Externalities. *International Regional Science Review*, *26*(2), 167–174.

Beauregard, R. A. (1998). Public-Private Partnerships as Historical Chameleons: The Case of the United States. In *Partnerships in Urban Governance* (pp. 52–70). New York: Palgrave Macmillan.

Beauregard, R. A. (2013). Shrinking Cities in the United States in Historical Perspective: A Research Note. In K. Pallagst, T. Wiechmann, & C. Martinez-Fernandez (Eds.), *Shrinking Cities: International Perspectives and Policy Implications*. London and New York: Routledge.

Bieri, D., & Kayanan, C. M. (2014, August 4). Improving TIF Transparency and Accountability: Towards a Consolidated View of TIF Activities in Michigan. *Social Science Research Network*, doi:10.2139/ssrn.2476143

Boyle, R., & Eisinger, P. (2001). The U.S. Empowerment Zone Program: The Evolution of a National Urban Program and the Failure of Local Implementation in Detroit, Michigan. *EURA Conference Paper*. Copenhagen.

Briffault, R. (2010). The Most Popular Tool: Tax Increment and Financing of Local Government the Political Economy. *Chicago Law Review*, *77*(1), 65–95.

Chapple, K., Markusen, A., Schrock, G., Yamamoto, D., & Yu, P. (2004). Gauging Metropolitan "High-tech" and "I-tech" Activity. *Economic Development Quarterly*, 18(1), 10–29.

Clark, T., Lloyd, R., Wong, K. K., & Jain, P. (2002). Amenities Drive Urban Growth. *Journal of Urban Affairs*, *24*(5), 493–515. doi:10.1111/1467-9906.00134

Coleman, D., & Murphy, B. (2014, May). *Economic Development*. St Louis: Better Together. https://www.bettertogetherstl.com/wp-content/uploads/2014/05/Better-Together-Economic-Development-Report-FULL-REPORT.pdf

Cortex Innovation Community (2020). Cortex Innovation Community 2020. Infographic. Retrieved May 19, 2023, from https://cortexstlorg.blob.core.windows.net/media/1486/cx_2020_cortex_infographic_v2-6.pdf

Dotzour, M., Grissom, T., Liu, C., & Pearson, T. (1990). Highest and Best Use: The Evolving Paradigm. *Journal of Real Estate Research*, 5(1), 17–32.

Drucker, J., Kayanan, C. M., & Renski, H. (2019). Innovation Districts as a Strategy for Urban Economic Development: A Comparison of Four Cases. *Social Science Research Network*, doi:10.2139/ssrn.3498319

Dye, R. F., & Merriman, D. F. (2006). Tax Increment Financing: A Tool for Local Economic Development. *Lincoln Land Lines*, *18*(1), 3–7.

Eggert, D. (2022, July 14). Stephen Ross Personally Lobbied Michigan Lawmakers for $100 Million Detroit Center Payout. *Crain's Detroit Business*. https://www.crainsdetroit.com/government/how-stephen-ross-got-100-million-detroit-center-innovation

Eisenschitz, A. (2010). Neo-liberalism and the Future of Place Marketing. *Place Branding and Public Diplomacy*, *6*(2), 79–86. doi:10.1057/pb.2010.12

Feldman, M. (1994). *The Geography of Innovation*. Netherlands: Kluwer Academic Publishers.

Feldt, B. (2018, April 13). Developer Plans More Buildings in St. Louis' Booming Cortex Innovation District. *St. Louis Post Dispatch*.

Florida, R. (2002). *The Rise of the Creative Class: And How It's Transforming Work, Leisure, Community, and Everyday Life*. New York: Basic Books.

Glaeser, E. L. (2011). *Triumph of the City: How Our Greatest Invention Makes Us Richer, Smarter, Greener, Healthier, and Happier*. New York: Penguin Books.

Gordon, C. (2008). *Mapping Decline: St. Louis and the Fate of the American City*. Philadelphia, PA: University of Pennsylvania Press.

Gramlich, E. (2022, January). *A Critical Explanation of Opportunity Zones*. Washington, DC: National Low Income Housing Coalition. https://nlihc.org/sites/default/files/A_Critical_Explanation_of_Opportunity_Zones.pdf

Grodach, C., & Loukaitou-Sideris, A. (2007). Cultural Development Strategies and Urban Revitalization: A Survey of US Cities. *International Journal of Cultural Policy, 13*(4), 349–370.

Gross, J. S. (2005). Business Improvement Districts in New York City's Low-Income and High-Income Neighborhoods. *Economic Development Quarterly, 19*(2), 174–189.

Hall, P. (1982). Enterprise Zones: A Justification. *International Journal of Urban and Regional Research1*, *6*(3), 416–421.

Harvey, D. (1989). From Managerialism to Entrepreneurialism: The Transformation in Urban Governance in Late Capitalism. *Geografiska Annaler. Series B, Human Geography, 71*(1), 3–17.

Houstoun, L. O. (2003). *Business Improvement Districts* (2nd ed). Washington, DC: Urban Land Institute.

Jessop, B., Brenner, N., & Jones, M. S. (2008). Theorizing Sociospatial Relations. *Environment and Planning D: Society and Space*, *26*(3), 389–401. doi:10.1068/d9107

Katz, B., & Wagner, J. (2014). The Rise of Innovation Districts: A New Geography of Innovation in America. *Metropolitan Policy Program at Brookings, May*. https://www.brookings.edu/essay/rise-of-innovation-districts/

Katz, B., Geolas, B., & Wagner, J. (2021). How Innovation Districts Can Help Drive an Inclusive Recovery. *Global Institute on Innovation Districts*. Retrieved July 27, 2022, from https://www.giid.org/how-innovation-districts-can-help-drive-an-inclusive-recovery/

Kayanan, C. M. (2022). A Critique of Innovation Districts: Entrepreneurial Living and the Burden of Shouldering Urban Development. *Environment and Planning A*, *54*(1), 50–66. doi:10.1177/0308518X211049445

Kayanan, C. M., Drucker, J., & Renski, H. (2022). Innovation Districts and Community Building: An Effective Strategy for Community Economic Development? *Economic Development Quarterly*, *36*(4), 343–354.

Klingmann, A. (2007). *Brandscapes: Architecture in the Experience Economy*. Cambridge, MA and London: MIT Press.

Krugman, P. R. (1991). Increasing Returns and Economic Geography. *Journal of Political Economy*, *99*(3), 483–499.

Lester, T. W. (2014). Does Chicago's Tax Increment Financing (TIF) Programme Pass the "But-for" Test? Job Creation and Economic Development Impacts Using Time-series Data. *Urban Studies*, *51*(4), 655–674. doi:10.1177/0042098013492228

Levy, P. R. (2001). Paying for the Public Life. *Economic Development Quarterly*, *15*(2), 124–131. doi:10.1177/089124240101500202

Lloyd, R. (2008). Neo-Bohemia: Art and Neighborhood Redevelopment in Chicago. *Cities and Society*, *24*(5), 215–229. doi:10.1002/9780470752814.ch16

Location. (n.d.). HUB RTP. Retrieved June 17, 2022 from: https://hub.rtp.org/location/

Logan, J. R., & Molotch, H. (2007). *Urban Fortunes: The Political Economy of Place*. Berkeley, CA: University of California Press.

Loukaitou-Sideris, A. (2000). Revisiting Inner-city Strips: A Framework for Community and Economic Development. *Economic Development Quarterly*, *14*, 165–181.

Malecki, E. J. (2010). Everywhere? The Geography of Knowledge. *Journal of Regional Science*, *50*(1), 493–513. doi:10.1111/j.1467-9787.2009.00640.x

Mallach, A., Haase, A., & Kattori, K. (2017). The Shrinking City in Comparative Perspective: Contrasting Dynamics and Responses to Urban Shrinkage. *Cities*, *69*(September), 102–108.

Martin, D. (2016, January 14). Nobody Calls It the "Innovation District" Anymore—Even the Mayor. *BostInno*, 7–9.

McMorrow, P. (2012, December 25). Seaport Is Rising, But Not from Tech. *The Boston Globe*, A.19.

McMorrow, P. (2014, May). Mayor Marty Walsh is Cleaning House. *Boston Magazine*.

Michigan Central (2022, February 4). Michigan Central Advances Plans for Mobility Innovation District. *Michigan Central*, https://michigancentral.com/michigan-central-advances-plans-for-mobility-innovation-district/

Mitchell, J. (2001). Business Improvement Districts and the "New" Revitalization of Downtown. *Economic Development Quarterly*, *15*(2), 115–123.

Mollenkopf, J. H. (1983). *The Contested City*. Princeton, NJ: Princeton University Press.

Neumann, T. (2016). *Remaking the Rust Belt: The Postindustrial Transformation of North America*. Philadelphia: University of Pennsylvania Press.

Pacewicz, J. (2013). Tax Increment Financing, Economic Development Professionals and the Financialization of Urban Politics. *Socio-Economic Review*, *11*(3), 413–440. doi.org/10.1093/ser/mws019

Pike, A. (2018, March 2). The Limits of City Centrism? We Need to Rethink How We Approach Urban and Regional Development. *British Politics and Policy at LSE*. https://blogs.lse.ac.uk/politicsandpolicy/the-limits-of-city-centrism/

Porter, M. E. (1990). *The Competitive Advantage of Nations*. New York: Free Press.

Rohe, W. M. (2012). *The Research Triangle: From Tobacco Road to Global Prominence*. Philadelphia, PA: University of Pennsylvania Press.

Sagalyn, L. B. (2007). Public/Private Development: Lessons from History, Research, and Practice. *Journal of the American Planning Association*, *73*(1), 7–22.

Sassen, S. (2001). *The Global City: New York, London, Tokyo*. Princeton, NJ: Princeton University Press.

Scott, A. J. (2001). Globalization and the Rise of the Entrepreneurial Economy. *European Planning Studies*, *9*(7), 813–826. doi:10.1080/0965431012007978

Seaport District, Boston, MA. (2022). realtor.com. Retrieved July 28, 2022, from: https://www.realtor.com/realestateandhomes-search/Seaport-District_Boston_MA/overview

Seidman, K. F. (2004, September). *Revitalizing Commerce for American Cities: A Practitioner's Guide to Urban Main Street Programs*. Washington, D.C.: Fannie Mae Foundation.

Shearmur, R., Carrincazeaux, C., & Doloreux, D. (Eds.). (2016). *Handbook on the Geographies of Innovation*. Northampton, MA: Edward Elgar Publishing.

Smith, A. G. (2017). *2017 Greater St. Louis Venture Capital Overview*. St. Louis, MO: Greater St. Louis Regional Chamber.

St. Louis Innovation District Tax Increment Financing (TIF) Redevelopment Plan. (2012). St. Louis, MO.

Storper, M., & Venables, A. J. (2004). Buzz: Face-to-face Contact and the Urban Economy. *Journal of Economic Geography*, *4*(4), 351–370. doi:10.1093/jnlecg/lbh027

Sugrue, T. (1996). *The Origins of the Urban Crisis*. Princeton, NJ: Princeton University Press.

Swanstrom, T. (2016). The Incompleteness of Comprehensive Community Revitalization. In J. DeFilippis (Ed.), *Urban Policy in the Time of Obama* (pp. 211–230). Minneapolis, MN: University of Minnesota.

Taft, C. (2018). Deindustrialization and the Postindustrial City, 1950–Present. *Oxford Research Encyclopedias* (June). doi:10.1093/acrefore/9780199329175.013.574

Tighe, J. R., & Ganning, J. P. (2015). The Divergent City: Unequal and Uneven Development in St. Louis. *Urban Geography2*, *36*(5), 654–673.

Weber, R. (2002). Extracting Value from the City : Neoliberalism and Urban Redevelopment. *Antipode*, *34*(3), 519–540.

Weber, R., & O'Neill-Kohl, S. (2013). The Historical Roots of Tax Increment Financing, or How Real Estate Consultants Kept Urban Renewal Alive. *Economic Development Quarterly*, *27*(3), 193–207. doi:10.1177/0891242413487018

Zandiatashbar, A., & Kayanan, C. M. (2020). Negative Consequences of Innovation-igniting Urban Developments: Empirical Evidence from Three US Cities. *Urban Planning*, *5*(3), 378–391.

Chapter 10

Ashton, P., Doussard, M., & Weber, R. (2012, July). The Financial Engineering of Infrastructure Privatization. *Journal of the American Planning Association*, *78*(3), 300–312.

Friedman, S. (2005). *Developer Solicitation for Publicly Owned Land*. International City/County Management Association. https://icma.org/sites/default/files/3815_.pdf

Friedman, S. B. (Ed.) (2016) *Successful Public/Private Partnerships: From Principles to Practices*. Urban Land Institute: Washington DC.

Malizia, E. (2012). Developing SynergiCity: The Real Estate Development Perspective. In P. Hardin & P. J. Armstrong (Eds.), *SynergiCity: Reinventing the Postindustrial City* (pp. 152–170). Chicago, IL: University of Illinois Press.

Sagalyn, L. (2012). Public-Private Engagement: P3 Promise and Practice. In B. Sanyal et al. (Eds.), *Planning Ideas that Matter: Livability, Territoriality, Governance, and Reflective Practice* (pp. 233–257). Cambridge, MA: MIT Press.

Navy-Focused Sources:

Baldor, L. (2021, November 2). Biden Taps Navy Admiral to be Joint Chief's Vice. *Associated Press*. Retrieved November 16, 2021, from: https://www.msn.com/en-us/news/us/biden-taps-navy-admiral-to-be-joint-chiefs-vice-chairman/ar-AAQf9Yi?ocid=uxbndlbing

Cocke, Sophie. (2014, January 30). Red Hill: EPA May Force New Fuel Leak Detection System for Toxic Spills. *Honolulu Civil Beat News*. Retrieved November 16, 2021, from: https://www.civilbeat.org/2014/01/red-hill-epa-may-force-new-fuel-leak-detection-system-for-toxic-spills/

Congressional Research Services (CRS). (2021, October 7). *Renewed Great Power Competition: Implications for Defense—Issues for Congress*. R43838. Retrieved November 16, 2021, from: https://crsreports.congress.gov/product/pdf/R/R43838

Defense.Gov. No author listed. (2020, February 10). *IMMEDIATE RELEASE; DOD Releases Fiscal Year 2021 Budget Proposal*. Retrieved November 16, 2021, from: www.defense.gov/News/Releases/Release/Article/2079489/dod-releases-fiscal-year-2021-budget-proposal/

Defense.Gov. Office of the Under Secretary of Defense (COMPTROLLER)/Chief Financial Officer. (2021, August). *Operation and Maintenance Overview United States Department of Defense Fiscal Year 2022 Budget Request*. Retrieved December 8, 2022, from: https://comptroller.defense.gov/Portals/45/Documents/defbudget/FY2022/FY2022_OM_Overview.pdf

EUL Fact Sheet (2014, September 14). Naval Facilities Engineer Command, eul_fact_sheet.pdf

GAO 15–346. Government Accountability Office. *Underutilized Facilities*. June 2015.

GAO 11–574. Government Accountability Office. *The Enhanced Use Lease Program Requires Management Attention.* June 2011.

Harper, J. (2020, October 6). Esper Calls for 500 Ship Navy to Counter China. *National Defense Magazine*. Retrieved November 16, 2021, from: https://www.nationaldefensemagazine.org/articles/2020/10/6/esper-calls-for-500-ship-navy-to-counter-china

McLeary, P. (2020, December 3). CJCS Milley Predicts DoD Budget "Bloodletting" To Fund Navy. *Breaking Defense Magazine*. Retrieved November 16, 2021, from: https://breakingdefense.com/2020/12/cjcs-milley-predicts-dod-bloodletting-to-fund-navy-priorities/

Principi, A. (2015, September 3). Time for a New BRAC. *The Hill News*. Retrieved November 16, 2021, from: https://thehill.com/blogs/congress-blog/homeland-security/252594-time-for-a-new-brac/

US Encyclopedia of Law (2014, September 14). United States Code 10-2667 (10-U.S.C.-2667). Leases: Non-Excess Property of Military Departments and Defense Agencies. Retrieved November 16, 2021, from: https://uscode.lawi.us/10-usc-2667/

Unified Facilities Criteria (UFC), WBDG, Whole Building Design Guide. Retrieved November 16, 2021, from: https://wbdg.org/FFC/dod/unified-facilities-criteria-ufc

Additional Bookmarks for UFC Below Due to Unconventional Organization

i. UFC 2-000-05N 900 Series Real Estate (Public Copy)
ii. Non-Government Standards Access | WBDG—Whole Building Design Guide
iii. UFC 1-201-02 Assessment of Existing Facilities for Use in Military Operations (wbdg.org)
iv. UFC 1-201-01 Non-Permanent DoD Facilities in Support of Military Operations (wbdg.org)
v. UFC 2-100-01 Installation Master Planning (wbdg.org)
vi. UFC 4-010-01 DoD Minimum Antiterrorism Standards for Buildings, with Change 1 (wbdg.org)

Other Sources:

Bruegmann, R. (2006). *Sprawl: A Compact History*. Chicago, IL: University Of Chicago Press.

Duany, A., Plater-Zyberk, E., & Speck, J. (2010). *Suburban Nation: The Rise of Sprawl and the Decline of the American Dream* (10th ed.). Albany, CA: North Point Press.

Dunham-Jones, E., & Williamson, J. (2008). *Retrofitting Suburbia: Urban Design Solutions for Redesigning Suburbs*. New York: Wiley.

Howell, D. D. (2015). *Army Installations of the Future: Urban + Shrinkage + Landscape*. Cambridge, MA: MIT Press.

Kim, J. (2021). *Bicycle Needs Assessment for Navy Installations.* Internal Navy Report, 2021.

Lang, J. (2015). *Urban Design: A Typology of Procedures and Products* (2nd ed.). London and New York: Routledge.

Lexicon of New Urbanism. Retrieved November 16, 2021, from: https://www.dpz.com/wp-content/uploads/2017/06/Lexicon-2014.pdf

Millichap. M. & Millichap Real Estate Investment Services. Six Reports: *2022 Office Outlook, 2022 Retail Outlook, 2022 Multi-family Outlook, 2022 Office Full Report, 2022*

Retail Research, and *2022 Multi-family Research*, https://www.marcusmillichap.com. Accessed from February 21 to March 27, 2022.

Stueuteville, R. (2019, January 29). *The Once and Future Neighborhood.* Retrieved November 16, 2021, from: https://www.cnu.org/publicsquare/2019/01/29/once-and-future-neighborhood

Talen, E. (2019). *Neighborhoods*. Oxford: Oxford University Press.

Zoning Overlay for NAS Oceana. City of Virginia Beach, Zoning Codes. Retrieved November 16, 2021, from: https://www.vbgov.com/government/departments/planning/areaplans/Pages/AICUZInformation.aspx.

Zoning General Land Use Plan NAS Oceana. City of Virginia Beach, Comp Plan. Retrieved November 16, 2021, from: https://www.vbgov.com/government/departments/planning/areaplans/Documents/Oceana/APZ1CZMASTERPLAN.pdf

USAF EUL Handbook. (2016, August). United States Air Force Civil Engineering Corps. Retrieved February 21, 2022, from: https://www.afcec.af.mil/Portals/17/documents/EUL/AF%20EUL%20Playbook%20-%2020160829.pdf?ver=2016-10-06-110839-517

US Census Bureau (n.d.). Quick Facts, Virginia Beach City, Virginia. Population, Census, April 1, 2020. https://www.census.gov/quickfacts/fact/table/virginiabeachcityvirginia/POP010220#POP010220

Index

accommodation and hospitality funds 44
affordable housing 4, 23, 34, 78, 78, 110, 129
agglomeration 100
ATFP 138

BID (Business Improvement Districts) 22, 23, 97, 99, 175
Big Box 113, 121, 130
blight 18, 23, 43, 63, 102, 128, 175
BRAC 138, 142, 191
brownfields 10, 14, 15, 17, 19, 21, 23, 24, 25, 26, 27

Camperdown 31, 33, 87
Carolina Panthers 28, 30, 38
Centennial American Properties 31, 89, 91, 93
CERCLA 24
community policing 76
concession agreement 32
contract law 29
contracts 10, 11, 28–38, 136, 138, 139, 141
Cortex Innovation Community 101, 102, 187
CRA 10, 11, 12, 61, 73, 127, 128, 129, 131, 132, 133, 134, 135, 176

Detroit Innovation District 101, 107
Durham 101, 105, 106

easements 140
eminent domain 10, 14, 15, 17, 19, 21, 23, 25, 26, 27, 52, 89, 102, 177
enterprise zones 12, 95, 97, 98, 99, 188

federal acquisition regulations (FAR) 141
FILOT 30, 33, 44, 120, 124
financing districts 12, 43
Fluor Field 11, 84, 88–94, 185, 186

Greenville Commons 46
Greenville Drive 89–93, 186
Greenville Local Development Corporation 42, 45, 51, 56, 59

historic tax credits 45, 52, 59
HUB RTP 101, 104–107, 109, 188

inflationary TIF 19
Inland Empire 112
Installment Purchase Revenue Bonds 43, 56
intermodal terminal 114, 115, 117

Kelo v New London Conn. 177
key contractual provisions 29, 33, 34, 35
key fundamentals 6

land lease xi

MILCON 136
Minor League Baseball 11, 83, 84, 85, 86, 89, 185, 186
multi-county business parks 44

National Defense Authorization Act 141
net present value 78, 134
New Economy Initiative 107
New Market Tax Credits 11, 45, 56, 57, 59
NEXT Innovation 56

One City Plaza 54
Opportunity Zones 98, 110, 181, 188

Peace Center 48, 49, 52
PILOT 31, 33, 120, 121, 145, 147, 148, 149

Pittsburgh 98
Poe West 58, 59
PolicyMap 130, 135
postindustrial 96, 99, 101, 107, 189, 190
property tax 17, 18, 20, 23, 24, 36, 37, 44, 64, 68, 97

Redevelopment TIF 19, 79, 102, 189
Request for Interest 10, 142, 171, 173
revenue bonds 43, 44, 56
Riverplace 52, 53, 89
ROI 4, 62, 63

Seaport Innovation District 101, 103, 104, 109
slum 17, 18, 21, 23, 26, 62, 63, 76
SRM Funds 136
SSRC 120, 123, 124, 125
St Louis 101, 102, 109, 111, 187, 188, 189, 190
Synthetic TIF 43

TID 19, 20, 21, 22, 61, 62, 63, 64, 68, 70, 76, 80
TIF 6, 10, 12, 14, 15, 16, 17, 18, 19–22, 23, 25, 26, 27, 43, 44, 48, 49, 50, 52, 54, 56, 61, 62, 64, 68, 79, 92, 95, 97, 98, 99, 102, 108, 127, 129, 130, 131, 132, 133, 134, 183, 187, 188, 189
TK Elevator 36, 38
Truist Park 36

underutilized real estate 13, 136, 137, 139, 145, 147, 148, 149
United Facilities Criteria 142
urban renewal 4, 17, 26, 95, 96, 97, 190
Utility Tax Credit 120

Westin Poinsett Hotel 50

yield on cost 129, 135

Printed in the United States
by Baker & Taylor Publisher Services

Printed in the United States
by Baker & Taylor Publisher Services